Transients of Modern Power Electronics

Transients of Modern Power Electronics

Hua Bai

Kettering University, Michigan, USA

Chris Mi

University of Michigan–Dearborn, USA

A John Wiley & Sons, Ltd., Publication

This edition first published 2011
© 2011, John Wiley & Sons, Ltd

Registered office
John Wiley & Sons Ltd, The Atrium, Southern Gate, Chichester, West Sussex, PO19 8SQ, United Kingdom

For details of our global editorial offices, for customer services and for information about how to apply for permission to reuse the copyright material in this book please see our website at www.wiley.com.

The right of the author to be identified as the author of this work has been asserted in accordance with the Copyright, Designs and Patents Act 1988.

Library of Congress Cataloguing-in-Publication Data

Bai, Hua, 1980-
  Transients of modern power electronics / Hua Bai, Chris Mi.
      p. cm.
   Includes bibliographical references and index.
   ISBN 978-0-470-68664-5 (hardback)
  1. Power electronics. 2. Transients (Electricity) 3. Electric current converters–Design and construction. I. Mi, Chris. II. Title.
   TK7881.15B34 2011
   621.381′044–dc22

                                                                                                    2011009746

A catalogue record for this book is available from the British Library.

Print ISBN: 978-0-470-68664-5
ePDF ISBN: 978-1-119-97172-6
oBook ISBN: 978-1-119-97171-9
ePub ISBN: 978-1-119-97276-1
Mobi ISBN: 978-1-119-97277-8

Typeset in 10/12pt Times by Laserwords Private Limited, Chennai, India

Contents

About the Authors

Hua (Kevin) Bai received his BS and PhD degrees in Electrical Engineering from Tsinghua University, Beijing, China in 2002 and 2007, respectively. He was a post doctoral fellow from 2007 to 2009 and an assistant research scientist from 2009 to 2010 at the University of Michigan–Dearborn in the United States. He is currently an Assistant Professor in the Department of Electrical and Computer Engineering, Kettering University, Michigan. His research interest is in the dynamic processes and transient pulsed power phenomena of power electronic systems, including variable frequency motor drive systems, high-voltage and high-power DC–DC converters, renewable energy systems, and hybrid electric vehicles.

Dr. Chris Mi is an Associate Professor of Electrical and Computer Engineering and Director of DTE Power Electronics Laboratory at the University of Michigan–Dearborn, Michigan in the United States.

Dr. Mi has conducted extensive research in electric and hybrid vehicles and has published more than 100 articles and delivered more than 50 invited talks and keynote speeches, as well as serving as a panelist.

Dr. Mi is the recipient of the 2009 Distinguished Research Award of the University of Michigan–Dearborn, the 2007 SAE Environmental Excellence in Transportation (also know as E2T) Award for "Innovative Education and Training Program in Electric, Hybrid, and Fuel Cell Vehicles," the 2005 Distinguished Teaching Award of the University of Michigan–Dearborn, the IEEE Region 4 Outstanding Engineer Award, and the IEEE Southeastern Michigan Section Outstanding Professional Award. He is also the recipient of the National Innovation Award (1992) and the Government Special Allowance Award (1994) from the China Central Government. In December 2007, Dr. Mi became a member of the Eta Kappa Nu, the Electrical and Computer Engineering Honor Society, for being "a leader in education and an example of good moral character."

Dr. Mi holds BS and MS degrees from Northwestern Polytechnical University, Xi'an, China, and a PhD degree from the University of Toronto, Canada. He was the Chief Technical Officer of 1Power Solutions from 2008 to 2010 and worked with General Electric Company from 2000 to 2001. From 1988 to 1994, he was a member of the faculty of Northwestern Polytechnical University, and from 1994 to 1996 he was an Associate Professor and an Associate Chair in the Department of Automatic Control Systems, Xi'an Petroleum University, China.

Dr. Mi is the Associate Editor of *IEEE Transactions on Vehicular Technology*, Associate Editor of *IEEE Transactions on Power Electronics – Letters*, associate editor of the *Journal of Circuits, Systems, and Computers* (2007–2009); editorial board member of *International Journal of Electric and Hybrid Vehicles*; editorial board member of *IET Transactions on Electrical Systems in Transportation*; a Guest Editor of *IEEE Transactions on Vehicular Technology*, Special Issue on Vehicle Power and Propulsion (2009–2010), and Guest Editor of *International Journal of Power Electronics*, Special Issue on Vehicular Power Electronics and Motor Drives (2009–2010). He served as the Vice Chair (2006, 2007) and Chair (2008) of the IEEE Southeastern Michigan Section. He was the General Chair of the Fifth IEEE International Vehicle Power and Propulsion Conference held in Dearborn, MI, September 7–11, 2009. He has also served on the review panel for the National Science Foundation, the US Department of Energy (2006–2010), and the Natural Sciences and Engineering Research Council of Canada (2010).

Dr. Mi is one of the two Topic Coordinators for the 2011 IEEE International Future Energy Challenge Competition.

Preface

Power electronics is a major branch of electrical engineering. The past few decades have witnessed exponential growth due to emerging applications in electric power systems, alternative energy, and hybrid electric vehicles. However, a popular view among many engineers and scholars is that power electronics has matured. In many circumstances, particularly among those who have only a cursory understanding of power electronic systems, power electronics are regarded as black boxes which could be sourced from the market. System integration is interpreted as sourcing these boxes, connecting them to other components, assembling them into the system, and then testing the system in environments that approximate those expected in the application.

This situation exists for several reasons. One is that power electronics lacks an instructive theoretical framework and design methodology. This deficiency directly leads to the empirical, vague, and inaccurate popularizing of power electronics as a black box. Realistically, a power electronics course should be multidisciplinary and involve semiconductor physics, digital signal processing, controls, circuits, computers, mechanical design, thermal and electromagnetic phenomena, and other disciplines. Understanding power electronics requires comprehension from macroscopic perspectives and microscopic factors. However, most of us still stay in the macroscopic world of control, topology, and circuits. Thus, compared to other courses like power systems and high-voltage engineering, power electronics has the lowest knowledge threshold to enter and it is assumed to behave like a pure applied engineering or even technician's discipline. Empirical coefficients, unconvincing simulation, unsophisticated electrical and mechanical concepts, and extensive reliance on testing often guide the design of power electronics.

As a matter of fact, the development of power electronic technology finds its roots in the development of semiconductor technology. One generation of power electronic systems is accompanied with one generation of semiconductor devices. Lacking an understanding of the physics of power semiconductor devices leads to the absence of the research fundamentals. Therefore, laboratory research activities which only care about macroscopic performance and ignore semiconductor physics are often accompanied by many unexpected failures. The switching actions of semiconductor devices introduce many transient processes which can challenge the safe operation of the power electronic systems. Statistically, nearly 70% of power electronic system failures happen in the transient processes instead of the steady state operations. However, in the mainstream of power electronics developments, critical transient processes are ignored and analysis methods are limited to averaged, steady state behavior. Topology, efficiency, total harmonic distortion, and output voltage ripples are often addressed and the voltage spike, in-rush current, minimum pulse width, and so on, are ignored.

The authors, whose point of view is validated by past experiences, believe the only way to simultaneously reach high performance, high reliability, and high design accuracy is to combine the analysis of the macroscopic control and microscopic transient processes. In order to establish a precise and instructive theoretical framework, collecting data from a variety of power electronic topologies is the first step in developing a theoretical framework for integrating microscopic and macroscopic phenomena. The authors have been involved in the development of: (i) a 6000 V, 1.25 MW three-level inverter, (ii) a 10 kW bidirectional isolated DC–DC converter, (iii) 10 kW battery chargers for plug-in hybrid electric vehicles, and (iv) a SiC JFET-based inverter. In conducting these research and development activities, the conflict between reliability and performance, the balancing of steady state and transient processes, and the struggle between the macroscopic and microscopic worlds were repeated for each development activity. In the high-power or high-power-density applications, observing, comprehending, and solving those transient processes is one of the most important steps. This has stimulated the writing of this book, entitled *Transients of Modern Power Electronics*. The authors hope that the book will inspire students and engineers to comprehend both the microscopic and macroscopic aspects of power electronics.

Chapter 1 gives a brief introduction to the state of the art of power electronics development which will facilitate readers' understanding of the present need in this domain. In Chapters 2 and 3, the major transient processes are addressed. The power electronic system is presented as an energy loop, energy components, and energy control. Typical transient processes are detailed in Chapter 4 for power electronics associated with hybrid electric vehicles, in Chapter 5 for alternative energy, and in Chapter 6 for battery management systems. In Chapters 7 and 8, the dead-band effect, minimum pulse width, and calculating errors, all critical elements of power electronics design, are detailed. Finally, Chapter 9 discusses future trends.

Since this work is a bold attempt and the data samples are limited in number, and although the authors have many years of experience in this domain, mistakes are unavoidable. Also, the authors have proposed many novel concepts in this book; however, these might not yet be accurate and may need improvement. The authors welcome all feedback that can be used to improve the contents of the book in future editions.

This work has been greatly supported by State Key Laboratory of Control and Simulation of Power System and Generation Equipment in Tsinghua University, China, the Department of Electrical and Computer Engineering at the

University of Michigan–Dearborn, and the Department of Electrical and Computer Engineering at Kettering University. The authors are grateful to all those who helped to complete the book. In particular, a large portion of the material presented in this book is the result of many years of work by the authors as well as other members of the research group of Professor Chris Mi and Professor Hua Bai. The authors are grateful to the many dedicated staff and graduate students who have made enormous contributions and provided supporting material for this book. The authors would like to thank Mr. Mariano Filippa who helped proofread chapter 1 to 3 of this book.

The authors would like to acknowledge various sources which granted permission to use certain materials or figures in the book. Best efforts were made to obtain permission for the use of these materials. If any of these sources were missed, the authors apologize sincerely for that oversight, and will gratefully rectify the situation in future editions of the book if it is brought to the attention of the publisher.

The authors would like to acknowledge The MathWorks, Inc. and ANSYS for providing software and support for their studies.

The authors also owe a debt of gratitude to their families, who have given tremendous support and made sacrifices during the process of writing this book.

Finally, they are extremely grateful to John Wiley & Sons, Ltd and its editorial staff for the opportunity to publish this book and helping in all possible ways.

1

Power electronic devices, circuits, topology, and control

1.1 Power electronics

Power electronics is a branch of engineering that combines the generation, transformation, and distribution of electrical energy through electronic means. In 1974, W. Newell described power electronics as a combination of electrical engineering, electronics, and control theory, which has been widely accepted today [1].

Power electronics has merged into various residential, commercial, and industrial domains. Application of power electronics encompasses renewable energy, transportation, defense, communication, manufacturing, utilities, and appliances. In the renewable energy field, power electronics covers distributed generation, control of electric power quality, wind power generation, and solar energy conversion. Modern power electronics consists of the research and development of novel power electronic semiconductors, new topologies, and new control algorithms. Power electronics is an interdisciplinary subject that involves traditional electrical engineering, electromagnetics, microelectronics, control, thermal fluid dynamics, and computer science.

More specifically, research in power electronics includes but is not limited to:

1. Theory, manufacture, and application of power electronic semiconductor devices.

2. Power electronic circuits, devices, systems and their relevant modeling, simulation, and computer-aided design.

3. Prediction and improvement of system reliability.

Transients of Modern Power Electronics, First Edition. Hua Bai and Chris Mi.
© 2011 John Wiley & Sons, Ltd. Published 2011 by John Wiley & Sons, Ltd.

4. Motor drive design, traction, and automation control.

5. Techniques for electromagnetic design and measurement.

6. Power electronics-based flexible AC transmission systems (FACTSs).

7. Advanced control techniques.

The study of power semiconductor devices is the foundation of modern power electronics. It began with the introduction of thyristors in the late 1950s. Today there are several types of power semiconductor devices available for power electronics applications, including gate turn-off thyristors (GTOs), power Darlington transistors, power metal oxide semiconductor field effect transistors (MOSFETs), insulated-gate bipolar transistors (IGBTs), and integrated-gate commutated thyristors (IGCTs). Recently, new materials with wideband energy gaps, such as silicon carbide (SiC) and gallium arsenide (GaS), are leading the direction of next-generation power semiconductor devices.

With the development of computer science and control theory, power electronics began to be utilized for industrial applications, for instance in motor drive and traction applications. Various remarkable control algorithms, such as field-oriented control (FOC) and direct torque control (DTC), have been developed for induction motor drives and permanent magnet motor drives [2–5].

With the development of power electronic technology, especially the maturity of high-voltage and high-power semiconductors, power electronics began to play an active role in power systems, improving their performance, cost, and controllability. FACTS is a typical example of power electronics in power system applications. The static reactive-power compensator (STATCOM) can eliminate excessive reactive power in the system so as to make the local power system more robust, environmentally friendly, and flexible [6–8].

Power supply is another area for the most popular power electronics applications. Spanning a wide range of power ratings, from ultralow power of a few milliwatts to several megawatts, and from a few volts to more than a thousand volts, power supplies based on power electronics occupy a large amount of market share. DC–DC converters [9], DC–AC inverters [10], AC–DC rectifiers [11], and AC–AC cyclo-converters [12] are typical of this field. Research in these power electronic technologies helps diversify topologies and the control methods. Furthermore, all of these topologies can be mathematically described, modeled, and simulated. For example, in order to mitigate thermal generation by the switching losses in hard-switched converters, soft-switching techniques were developed where nearly all circuits have their own unique topology mathematically modeled according to their own operation modes [13–17]. Advanced control algorithms and diverse topologies can all be validated through the use of sophisticated analytical and numerical analysis tools, especially after the feasibility and accuracy of such tools have been validated widely in consumer and industrial applications.

1.2   The evolution of power device technology

Power semiconductors are the fundamental building blocks of power electronics. Each generation of semiconductors determines its corresponding generation of power electronic technology. The first power electronic device ever created was the mercury arc rectifier in 1900. The grid-controlled vacuum rectifier, ignitron, and thyratron followed later. These devices were found in numerous applications in industrial power control until 1950. At this time, the invention of the transistor in 1948 marked a revolution in the field of electronics. It also paved the way for the introduction of the silicon-controlled rectifier, announced by General Electric in 1957, commonly known nowadays as the thyristor.

All of these semiconductor devices can be classified as the following three types:

1. **Uncontrolled devices:** devices that do not need any trigger signals to control their on/off action, such as a rectifying diode.

2. **Semi-controlled devices:** devices that can be triggered on but cannot be turned off through control signals. A typical example is a thyristor, where the only way to turn it off is to reverse the polarity of the voltage across it and wait until the current reaches zero.

3. **Fully controlled devices:** also known as self-controlled devices, these devices can be turned on and off by the gate signals. Typical examples include bipolar junction transistors (BJTs), IGBTs, MOSFETs, GTOs, and IGCTs.

The common aspects of thyristors and GTOs are their high power ratings (most recently reaching over 6000 V/6000 A) and slow switching speed. They have always been the primary choice in high-voltage and high-power inverters (voltage source or current source inverters) until IGCTs emerged. Due to their slow switching speed, the switching frequency of thyristors and GTOs cannot be too high, otherwise a large switching loss will eventually damage the device. In medium-voltage applications, thyristors and GTOs have been replaced by high-voltage IGBTs or IGCTs. However, in high-voltage DC applications, thyristors and GTOs still dominate.

BJTs and MOSFETs were developed simultaneously in the late 1970s. BJTs are current-controlled devices while MOSFETs are voltage-controlled devices. Power BJTs have gradually been phased out while MOSFETs and IGBTs have become dominant in power electronics, especially in low- to medium-power applications. Compared to BJTs, MOSFETs can operate at higher switching frequencies while having lower switching losses. The only disadvantage of MOSFETs is their higher on-state voltage compared to that of IGBTs.

An IGBT is basically a combination of a BJT and a MOSFET [18]. It has been an important milestone in the history of power semiconductor devices. Its switching frequency can be much higher than that of BJTs, and its electrical capabilities are much higher than those of MOSFETs. Currently, IGBTs can reach 6000 V/600 A or 3500 V/1200 A. The operational details of IGBTs will be further explained in the next few chapters.

IGCTs were introduced by ABB in 1997 [19]. Presently they can reach 4500 V/4000 A. Essentially, a gate-controlled thyristor (GCT) is a four-layer thyristor, being simple to turn on but difficult to turn off. However, with the introduction of "integrated gates," the turn-off process is accelerated by shifting all the current from the GCT to its gate. Therefore, in the turn-off process, an IGCT behaves as a transistor. The advantages of IGCTs over GTOs include: faster switching, uniform temperature distribution within the junction, and snubber-free operation. One of the IGCT's disadvantages is that a short circuit is formed across its terminals during failure, which is not desirable in most power electronics applications.

Power semiconductor device development now extends beyond just semiconductor design. With the increase in various power electronics applications, more and more power devices tend to integrate gate-drive circuits, overcurrent protection, and other additional functions inside the module. Thus intelligent power modules (IPMs) have emerged for up to several hundred kilowatts for IGBTs [20]. IGCTs are typical IPMs. An IGCT integrates the gate with a GCT. Some types of IGCTs can even process self-diagnosis and feed back their status to the microcontroller.

The above semiconductors are silicon based. It is expected that in the future silicon devices will still keep their dominance. However, other materials have shown promise as well. For example, the silicon carbide (SiC) semiconductor has a wider bandgap (3.0 eV for 6H-SiC), higher saturation velocity (2×10^7 cm/s), higher thermal conductivity (3.3–4.9 W/cm K), lower on-state resistance (1 mΩ/cm^2), and higher breakdown electric field strength (2.4 MV/cm) [21]. Therefore, SiC-based power devices are expected to show superior performance compared to traditional silicon (Si) power switches. Since SiC power devices can operate at higher switching frequencies, the size of passive components (inductors and capacitors) can be reduced significantly in SiC-based power electronic converters. The associated heat sink size will also be reduced due to the lower losses compared to a conventional power electronic converter. Higher junction temperatures will result in much simpler cooling mechanisms. It is predicted that SiC devices will have a significant impact on the next generation of power electronic systems.

1.3 Power electronic circuit topology

Power semiconductor switches are the fundamental building blocks of power electronic converters. Switching actions are the core of power electronic converters.

1.3.1 Switching

Switches are the components responsible for controlling energy flow in power electronics. Consider an IGBT as an example. When the gate is supplied with a voltage higher than its turn-on threshold, the collector and emitter will show a low impedance between its terminals, therefore the device will have an "on state" equivalent to a closed switch. When the gate voltage is lower than its turn-on threshold, high impedance will be present between the collector and emitter, preventing current flow and transitioning to an "off state." Switching is the repetitive action of changing the semiconductor switch from an on state to an off state and vice versa. By controlling the state of the power switches, the energy flow is controlled through different paths.

Figure 1.1 shows a typical buck converter topology. The buck or step-down topology converts an input DC voltage to a lower output DC voltage. This is accomplished by controlling its main switching device, rerouting the flow of energy, and converting from a higher voltage to a lower voltage.

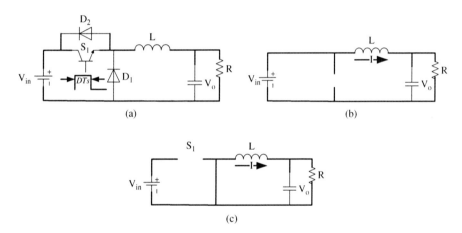

Figure 1.1   A buck circuit and its topology under different operation modes: (a) a buck converter, (b) topology when S_1 *is on, and (c) topology when* S_1 *is off.*

The buck converter consists of a primary power input V_{in}, an active switch S_1 (parallel diode D_2 is temporarily neglected), a clamping diode D_1, an inductor L, and an output capacitor holding voltage V_o which is assumed to be constant. All elements are shown in Figure 1.1a. R is added as a load to the output of the circuit. In this topology, D is the switch-on/off duty cycle and T_s is the switching period. When S_1 is on, the equivalent circuit is illustrated in Figure 1.1b where the current of the inductor increases linearly. The voltage drop across L is $V_{in} - V_o$. When S_1 is off, the equivalent circuit is shown in Figure 1.1c where the inductor current decreases linearly. Since current through the inductor cannot stop instantaneously, the inductor voltage will reverse its direction, therefore forcing the clamping diode to conduct. The voltage drop across L is now $-V_o$. By switching on and off alternately, the inductor average current is maintained and keeps up with the output power requirements.

For the purpose of analyzing this topology, the switching actions are assumed to take place in a negligible time interval. Therefore the switching process is divided into two independent states. Based on this premise, the buck converter shown in Figure 1.1a is mathematically modeled as follows.

When S_1 is on,

$$L\frac{di}{dt} = V_{in} - V_o \tag{1.1}$$

When S_1 is off,

$$L\frac{di}{dt} = 0 - V_o \tag{1.2}$$

It is observed that the switching action only provides a possible alternative for energy flow. It does not have to change the circuit topology and thereby the energy loop. For example, in Figure 1.1a, if, for some reason, the initial current in the inductor flows in the opposite direction, the current flow is always through D_2 regardless of the on or off state of S_1.

1.3.2   Basic switching cell

Figure 1.2 shows two basic switching cells defined in [22] a P cell and an N-cell. Each cell consists of one active switch, one diode and one current load; (+) stands for the positive DC-bus voltage and (−) represents the negative DC-bus voltage; (→) means current flows out of the bridge and (←) for current flowing into the bridge. For the P-cell, the active switching device is connected to the positive terminal (+). The cathode of the diode and the other node of the active switch are connected to the load. This is the opposite for the N-cell, which is shown as Figure 1.2b.

All power electronic converters are different combinations of the above basic cells. The reason lies in the theory of energy continuity. When an active switch is turned off, an alternative loop is needed to exhaust the excessive energy. Therefore an auxiliary diode should exist to commute the current. This diode is also called a freewheeling diode.

1.3.3   Circuit topology of power electronics

Power electronic circuits convert electrical energy from one form to another. In [23], a power electronic circuit is defined as "the part of a system that actually manipulates the flow of energy." It also provides "an interface between two other systems."

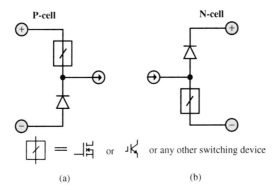

Figure 1.2   Basic switching cells: (a) a P-cell and (b) an N-cell. © [2009] IEEE. Reprinted, with permission, from IPEMC 2009.

In the theoretical analysis of power electronics, the components of power electronic systems and connections are considered lossless. With this assumption, circuit topologies and the theoretical investigation of the power electronic circuits can be greatly simplified.

In the domain of power electronics, topology stands for the specific position of each component and its electrical relationships. Junctions, loops, and simplified symbols of the components are the major elements of power electronics topologies.

Due to the existence of semiconductor switches, power electronics topologies have their own peculiarities, such as:

1. **Time variant:** the operation of the circuit is different depending on whether the switch is on or off as shown in Figure 1.1. Considering the configuration of Figure 1.1b as A and that of Figure 1.1c as B, as the time step to analyze the topology is reduced, some intermediate processes need to be taken into account, such as the switching process of the switches. Therefore a potential configuration C exists between A and B, where some other parameters, for example, the junction capacitance and stray inductance, should be considered. This will be illustrated in later chapters.

2. **Space variant:** if the wire connections are regarded as ideal conductors, the physical location of the circuit components has no effect on the circuit topology. However, in many specific applications where a more detailed analysis is needed, those connections cannot be regarded as ideal wires. Stray inductance and capacitance need to be included in the circuits.

In Figure 1.3, an H-bridge inverter is shown. Leg 1 comprises T_1 and T_2, and leg 2 comprises T_3 and T_4. The physical placement of these devices is shown in Figure 1.3b, where leg 2 is more distant away from the DC-bus capacitor than leg 1. Therefore the stray inductance of loop 2 is larger than loop 1.

Based on the above analysis, the electrical behavior is different when the parasitic/stray elements are included. There is also a big difference when the concept of topology is used in the investigation of transient processes, as defined in [24].

For the two bridges of three-level converters, the traditional topology assumes that the semiconductors, such as IGCTU1 and IGCTV1 in Figure 1.4, are subject to the same voltage stress in the switching-off process. However, in Figure 1.4, large amounts of stray inductances are present in the commutating process. They exist in the loop made of snubber diodes (Ls_1–Ls_6), clamping diodes, IGCTs, and so on. Due to the different configuration of the loops and different distances from the snubber circuit, the parasitic inductances of the commutating loop

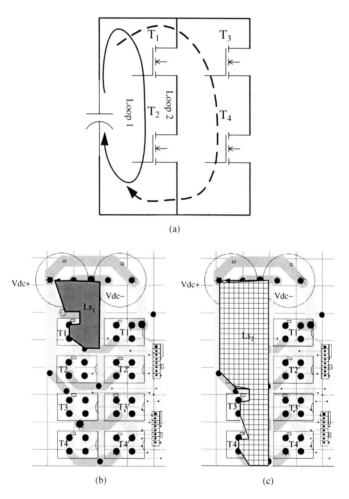

(a)

(b)   (c)

Figure 1.3   H-bridge loops and their stray inductance: (a) H-bridge, showing loop 1 and loop 2, (b) stray inductance of loop 1, (c) stray inductance of loop 2, and (d) finite-element-method analysis for loop 2.

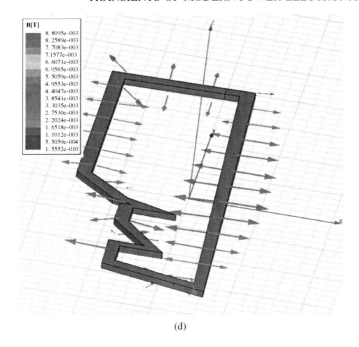

(d)

Figure 1.3   (continued)

for different IGCTs vary, in the case of figure shown, being $L_1 = 350\,\text{nH}$ and $L_2 = 267\,\text{nH}$. The reason why $L_2 < L_1$ is because the snubber circuit is closer to bridge V and more distant away from bridge U. Note that Figure 1.4 is only a schematic representation and does not show the real distribution of components and loops in three dimensions. The unequal distance to the snubber circuit generates the voltage spikes undertaken by IGCTs.

In order to achieve high-efficiency energy conversion, not only the control algorithm, but also the energy loop and the energy storage should be optimized. These parasitic loops will present some side effects, as illustrated in the following chapters.

1.4   Pulse-width modulation control

Circuit topology provides an effective way to analyze power electronic systems. The switches in the topology are controlled through triggering the gate signals which in the real world are typically digital pulses generated by microcontrollers with certain control algorithms. Among all the signal modulation schemes, pulse-width modulation (PWM) is the most popular strategy and involves modulation of the duty cycle to produce the required voltage, current, or power to the load [25]. Particularly in the domain of power supply and motor control, PWM plays a dominant role.

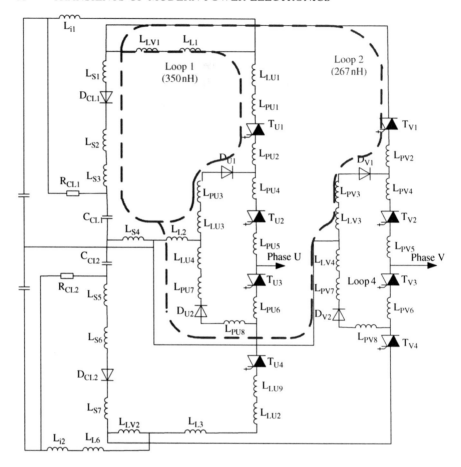

Figure 1.4   Transient commutating topology.

PWM utilizes a pulse sequence whose pulse width is varied over time or over different switching cycles, resulting in the variation of the average value of the waveform. Consider a square waveform as shown in Figure 1.5a, which has a minimum value y_{min}, a maximum value y_{max}, and a duty ratio D. The average value of the waveform is then

$$\bar{y} = \frac{1}{T} \left(\int_0^{DT} y_{max} \, dt + \int_{DT}^T y_{min} \, dt \right)$$

$$= \frac{DT y_{max} + (T - DT) y_{min}}{T} \tag{1.3}$$

$$= D y_{max} + (1 - D) y_{min}$$

Suppose $y_{min} = 0$ and $y_{max} = 1$; then Equation 1.3 turns out to be $\bar{y} = D$.

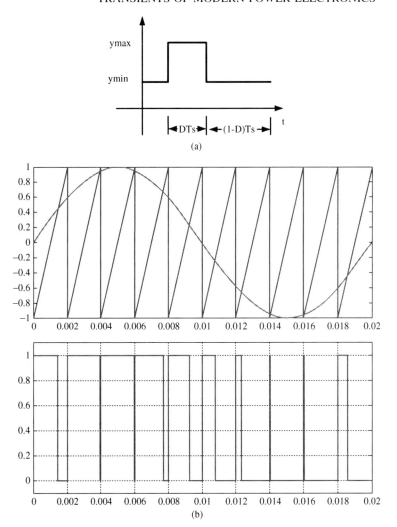

Figure 1.5   Illustration of PWM methods: (a) a signal with duty ratio D *and (b) PWM waveforms.*

The simplest way to generate a PWM signal is the sine-triangle PWM method, which adopts a sinusoidal waveform as the reference signal, a sawtooth or a triangular waveform as the carrier waveform, and a comparator. When the value of the reference signal is more than the carrier waveform, the PWM signal output will be in a high-state, otherwise it will be in a low-state, as shown in Figure 1.5b.

Delta modulation is another method of PWM control, where the output signal is restrained by two limits, that is, the upper limit and the lower limit. The offset between these two limits is a constant. Once the output signal reaches one of the limits, the PWM signal changes state, as shown in Figure 1.6. More details can be found in [26].

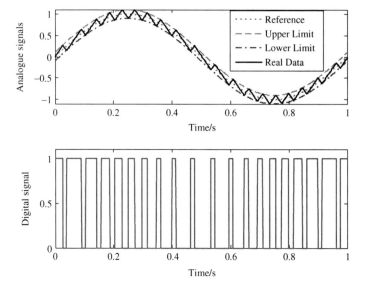

Figure 1.6   Delta PWM.

Sigma–delta is the third method of PWM control, shown in Figure 1.7, where the output signal is subtracted from a reference signal to form an error signal [27]. This error is integrated, and when the integral of the error exceeds the limits, the output changes state.

Space vector modulation is a PWM control algorithm for multi-phase AC wave generation, in which the reference signal is sampled regularly [28, 29]. After each sample, non-zero active switching vectors adjacent to the reference vector and one or more of the zero switching vectors are selected for the appropriate fraction of the sampling period in order to synthesize the reference signal [30]. The detailed theory of space vector PWM and its application in three-level DC–AC inverters will be explained in Chapters 4 and 5.

There are analogue integrated circuits (ICs) on the market that perform these PWM control methods, with low power and reduced component count as their main advantages. However, they lack flexibility in configurability. Many digital circuits (e.g., microcontrollers) are capable of generating PWM signals. They typically use a counter that increments periodically and is reset at the end of every period of the PWM. When the counter value reaches a configurable reference value, the PWM output changes state from high to low or vice versa. An example of a PWM-capable microcontroller is the TMS320F2XX from Texas Instruments (or TI).

When incrementing counters work in microcontrollers, the PWM method used is the intersecting method. The comparator function is performed by comparing

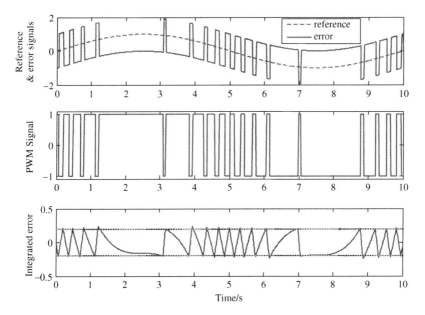

Figure 1.7   Sigma–delta PWM.

the current counter value to a reference value, both digitally. The duty cycle can no longer vary continuously due to the limited counter resolution. Therefore, the duty cycle varies in discrete steps. For example, if the maximum counter value is 256, the duty cycle resolution is 0.39%.

Three types of PWM are possible (Figure 1.8), whose difference lies only in the different sawtooth or triangular carrier signals applied to generate the PWM waveforms using the intersecting method:

1. Center-aligned PWM, where all PWM signals generated have their centers aligned.

2. Leading edge aligned, where the rising edges of all PWM signals generated are aligned.

3. Trailing edge aligned, where the falling edges of all PWMs are aligned.

High-frequency PWM output (PWM voltage) can be easily realized using power semiconductor switches. The PWM signals are used to control the state of the switches which directly determine the load voltage/current. The switches are either off (not conducting any current) or on (have no voltage drop across them with ideal switches). The product of current and voltage defines the instantaneous power dissipated inside the switch, thus no power is dissipated in an ideal switch [30]. However, semiconductor switches are always not ideal, although the losses of these devices are relatively small compared to the power they can deliver to the load. Additional control strategies, such as soft-switching control, can be

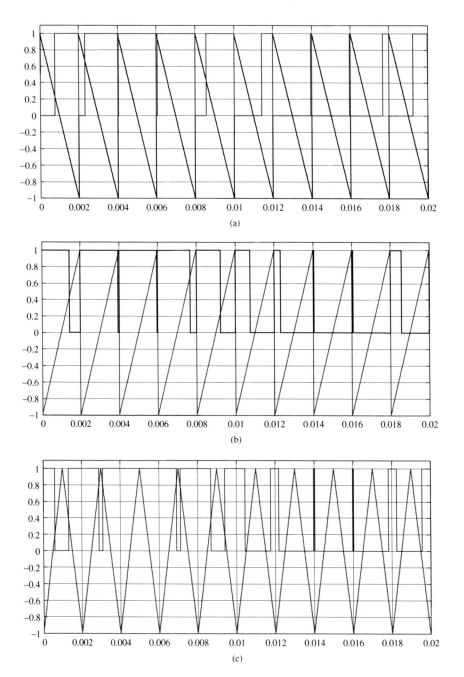

Figure 1.8   Three types of PWM signals: leading edge modulation, trailing edge modulation, and centered pulses (from top to bottom).

used to achieve very high efficiency for power converters, even at very high frequencies.

PWM is also often used to control the supply of electric power to another system. For example, induction motors used for pumps and blowers mostly employ a specific type of PWM control with a power converter. These power converters receive energy from the AC grid, rectify the AC input to a DC voltage, and then invert to a variable voltage, variable frequency (VVVF) AC voltage by means of PWM control [31]. Controlling the speed of the motor is done inside the microcontrollers by changing the modulation index and the frequency of the pulse sequences. By chopping the DC-bus voltage with the appropriate duty cycle, the output will be at the desired level. The voltage and current ripple are usually filtered with an inductor and/or a capacitor.

1.5 Typical power electronic converters and their applications

Power electronic converters are found in many applications where there is a need to change the form of electrical energy (i.e., voltage, current, or frequency). The ratings of power converters range from a few milliwatts (e.g., in a mobile/cell phone) to hundreds of megawatts (e.g., in a high-voltage DC transmission system). In contrast to electronic systems concerned with transmitting and processing signals and data which carry very large amounts of information with small amounts of energy, power electronics systems handle a large amount of electrical energy along with information. The power conversion systems can be classified according to the types of input and output:

- **AC to DC (rectifier):** converts AC input to DC output. This takes place in electronic devices and systems that use DC power as the power supply with only AC source available.

- **DC to AC (inverter):** converts DC input to AC output, with fixed or variable frequency and voltage. Typical applications are an uninterruptable power supply (UPS), emergency lighting, and motor control.

- **DC to DC (chopper):** converts a DC voltage to a different DC voltage. DC sources include batteries, photovoltaic arrays, and fuel cells. Buck, boost, buck–boost, Cuk circuit, half-bridge DC–DC, and full-bridge DC–DC are typical examples.

- **AC to AC (cyclo-converter):** converts one form of AC voltage to another AC voltage with a different frequency or voltage.

An AC–DC converter (rectifier) is typically found in many consumer electronic devices and residential applications, for example, cellular phones, televisions, computers, washers, dryers, air-conditioners, automobiles, and so on. The power range is typically from tens of watts to several hundred watts.

Variable speed drives (VSDs) are typical DC-AC converters which are used to control electric motors, including induction, DC, and permanent magnet motors. The power rating of VSDs varies from a few hundred watts to tens of megawatts.

It is worthwhile to point out that, in some applications, more than one conversion system is used. For example, battery chargers take the input voltage from the AC grid, rectify it to a DC voltage, and then convert the DC voltage to another DC voltage at a different level. Similarly, variable speed drive (VSD) systems also employ AC input, rectify it to DC, and convert back to another AC with different frequency and voltage to drive AC motors.

1.6   Transient processes in power electronics and book organization

Transient processes in power electronic converters occur in multiple forms. The common aspects are the short-timescale and large-energy exchange in these transient processes. With this in mind, Chapter 2 will categorize the research on power electronics into the macroscopic and microscopic domains. The perspective of this book is microscopic research in power electronics, and the main focus of the following chapters will range from devices and topology to circuits and systems.

In Chapter 3, the transients in power electronic devices are described. Three of the most widely used devices including Si IGBT, Si IGCT, and SiC JFET, are modeled.

As one of the most interesting applications, electric vehicles (EVs) and hybrid electric vehicles (HEVs) have attracted a lot of attention lately. Chapter 4 will discuss the application of power electronics in EVs and HEVs. An inverter-fed drive system and battery chargers will be detailed.

In Chapter 5, alternative energies with power electronics will be addressed. Solar energy, wind energy, and fuel cells are the three main subjects. Power electronic converters are needed to control their variable output and provide high-efficiency energy. The critical transient processes in these applications will be analyzed.

As another typical example, the battery management system in the HEV and plug-in HEV (PHEV) will be addressed in Chapter 6. This system is highly related to power electronics to balance the energy distribution among the battery cells in the same string. The energy flow in the battery management system will be described.

While Chapter 2 analyzes the transients at the device level, Chapters 4–6 study them at the system level. Chapters 7 and 8 will combine the device-level with the system-level perspectives. The influence of three typical microscopic factors, dead band, minimum pulse width, and modulated error in power electronic systems, will be demonstrated in detail.

Finally, chapter 9 looks at the future trends of power electronics, including devices, topology, packaging, systems and applications. The impact of new devices and topology on the transients of power electronics are discussed.

References

1. Newell, W.E. (1974) Power electronics – emerging from Limbo. *IEEE Transactions on Industry Applications*, **1A–10** (1), 7–11.

2. Meng, M. (2007) Voltage vector controller for rotor field-oriented control of induction motor based on motional electromotive force. Industrial Electronics and Applications, ICIEA 2007, pp. 1531–1534.

3. Juhasz, G., Halasz, S., and Veszpremi, K. (2000) New aspects of a direct torque controlled induction motor drive. Proceedings of IEEE International Conference on Industrial Technology, pp. 43–48.

4. Idris, N.R.N. and Yatim, A.H.M. (2004) Direct torque control of induction machines with constant switching frequency and reduced torque ripple. *IEEE Transactions on Industrial Electronics*, **51** (4), 758–767.

5. Lee, K.-B., Song, J.-H., Choy, I., and Yoo, J.-Y. (2002) Torque ripple reduction in DTC of induction motor driven by three-level inverter with low switching frequency. *IEEE Transactions on Power Electronics*, **17** (2), 255–264.

6. Xi, Z., Parkhideh, B., and Bhattacharya, S. (2008) Improving distribution system performance with integrated STATCOM and super-capacitor energy storage system. Power Electronics Specialists Conference, pp. 1390–1395.

7. Khederzadeh, M. (2007) Coordination control of STATCOM and ULTC of power transformers. Universities Power Engineering Conference, pp. 613–618.

8. Chen, J., Song, S., and Wang, Z. (2006) Analysis and implement of thyristor-based STATCOM. International Conference on Power System Technology, pp. 1–5.

9. Mi, C., Bai, H., Wang, C., and Gargies, S. (2008) The operation, design, and control of dual H-bridge based isolated bidirectional DC-DC converter. *IET Power Electronics*, **1** (3), 176–187.

10. Zhang, W. and Chen, W. (2009) Research on voltage-source PWM inverter based on state analysis method. International Conference on Mechatronics and Automation, pp. 2183–2187.

11. Alves, R.L. and Barbi, I. (2009) Analysis and implementation of a hybrid high-power-factor three-phase unidirectional rectifier. *IEEE Transactions on Power Electronics*, **24** (3), 632–640.

12. Sugimura, H., Eid, A.M., Kown, S.-K., Lee, H.W., Hiraki, E., and Nakaoka, M. (2005) High frequency cyclo-converter using one-chip reverse blocking IGBT based bidirectional power switches. International Conference on Electrical Machines and Systems, pp. 1095–1100.

13. Zhu, L. (2006) A novel soft-commutating isolated boost full-bridge ZVS-PWM DC-DC converter for bidirectional high power application. *IEEE Transactions on Power Electronics*, **21**, 422–429.

14. Li, H., Peng, F.Z., and Lawler, J.S. (2003) A natural ZVS medium-power bidirectional DC-DC converter with minimum number of devices. *IEEE Transactions on Industrial Applications*, **39**, 525–535.

15. Li, H. and Peng, F.Z. (2004) Modeling of a new ZVS bi-directional DC-DC converter. *IEEE Transactions on Aerospace and Electronic Systems*, **40** (1), 272–283.

16. Lin, B.R. and Shih, K.L. (2010) Analysis and implementation of a soft switching converter with reduced switch count. *IET Power Electronics*, **3** (4), 559–570.

17. Zhang, H., Wang, Q., Chu, E. *et al.* (2010) Analysis and implementation of a passive lossless soft-switching snubber for PWM inverters. *IEEE Transactions on Power Electronics*, **26** (2), 411–426.

18. Khanna, V. (2005) Power device evolution and the advent of IGBT, in *Insulated Gate Bipolar Transistor IGBT Theory and Design*, John Wiley & Sons, Inc., Hoboken, NJ, pp. 1–33.

19. Steimer, P.K., Gruning, H.E., Werninger, J., Carroll, E., Klaka, S., and Linder, S. (1997) IGCT – a new emerging technology for high power, low cost inverters. Industry Applications Conference, pp. 1592–1599.

20. Motto, E., Donlon, J., Shang, M., Kuriaki, K., Iwagami, T., Kawafuji, H., and Nakano, T. (2008) Large package transfer molded DIP-IPM. Industry Applications Conference, pp. 1–5.

21. Mazzola, M.S. and Kelley, R. (2009) Application of a normally OFF silicon carbide power JFET in a photovoltaic inverter. Applied Power Electronics Conference and Exposition, pp. 649–662.

22. Tolbert, L.M., Peng, F.Z., Khan, F.H., and Li, S. (2009) Switching cells and their implications for power electronic circuits. International Power Electronics and Motion Control Conference, pp. 773–779.

23. Rosado, S., Prasai, A., Wang, F., and Boroyevich, D. (2005) Study of the energy flow characteristics in power electronic conversion systems. Electric Ship Technologies Symposium, pp. 333–339.

24. Bai, H., Zhao, Z., and Mi, C. (2009) Framework and research methodology of short-timescale pulsed power phenomena in high-voltage and high-power converters. *IEEE Transactions on Industrial Electronics*, **56** (3), 805–881.

25. Wu, C.-C. and Young, C.-M. (1999) A new PWM control strategy for the buck converter. Conference of the IEEE Industrial Electronics Society, pp. 157–162.

26. Flood, J.E. and Hawksford, M.J. (1971) Exact model for delta-modulation processes. *Proceedings of the Institution of Electrical Engineers*, **118** (9), 1155–1161.

27. Zierhofer, C.M. (2008) Frequency modulation and first-order delta sigma modulation: signal representation with unity weight Dirac impulses. *IEEE Signal Proceedings Letters*, **15**, 825–828.

28. Shun, J. and Yanru, Z. (2004) A novel three-level SVPWM algorithm considering neutral-point control, narrow-pulse elimination and dead-time compensation. International Power Electronics and Motion Control Conference, Vol. 2 (1), pp. 688–693.

29. Liu, H.L. and Cho, G.H. (1994) Three-level space vector PWM in low index modulation region avoiding narrow pulse problem. *IEEE Transactions on Power Electronics*, **9** (5), 481–486.

30. Pulse-width modulation. http://en.wikipedia.org/wiki/Pulse-width_modulation (accessed March 23, 2011).

31. Chan, W.L., Suen, S.M., and So, A.T.P. (1997) A study on electrical performance of modern VVVF drives for HVAC applications. International Conference on Advances in Power System Control, Operation and Management, pp. 825–830.

2

Macroscopic and microscopic factors in power electronic systems

2.1 Introduction

In Chapter 1, the power electronic circuit shown in Figure 1.1 and Equations 1.1 and 1.2 simplify the system by neglecting the actual switching process, which in fact takes some time to occur. Present semiconductor switches are still far from ideal and the switching process is not negligible. Specifically, the disadvantages of this simplified model and analysis are as follows.

First, the ideal model is unable to calculate the switching loss, which is an important steady-state parameter. At the on/off switching transition, there is always a region where neither current nor voltage is zero. Mathematically we can define the integration of the product of current and voltage across the switch in the switching process as the lost energy inside the switch as shown in Figure 2.1. Switching loss [1, 2], defined by the switching energy averaged in one switching period, characterizes the switch performance. The unit for switching energy loss is Joule (J) and that for switching power loss is Watt (W).

In Figure 2.1, the turn-off loss is defined as

$$P_{turn\text{-}off} = \frac{1}{T_s} \int_{t_1}^{t_2} v(t) \times i(t) dt \qquad (2.1)$$

Transients of Modern Power Electronics, First Edition. Hua Bai and Chris Mi.
© 2011 John Wiley & Sons, Ltd. Published 2011 by John Wiley & Sons, Ltd.

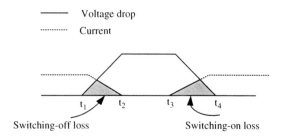

Figure 2.1 Switching loss with simplified and linearized switching characteristics.

and the turn-on loss is

$$P_{turn\text{-}on} = \frac{1}{T_s} \int_{t_3}^{t_4} v(t) \times i(t)dt \qquad (2.2)$$

Other switching losses include reverse recovery loss [3] and forward recovery loss [4], which will be described in later chapters. The above definition facilitates the loss calculation to assess the thermal characteristics of the system but does not explain the mechanisms of switching losses. In the later sections we will explain that switching losses are of an electromagnetic nature and could be estimated accurately via Maxwell's equations [5].

In order to estimate the switching losses, the current and voltage waveforms are linearized at different moments in the switching process. This calculation method has been widely accepted by power electronics engineers.

From Equations 2.1 and 2.2, high-switching-frequency, high-voltage, and high-current applications generate higher switching losses, which will cause a higher temperature-rise in the devices and therefore require a larger heat sink. Various strategies are proposed to handle the high-frequency or high-power applications, for example, soft-switching control and snubber circuits.

Second, the ideal model conceals the influence of vital short-timescale parameters. For example, all the influences of stray inductance and reverse recovery characteristics of diodes are not covered by Figure 1.1 and Equations 1.1 and 1.2. No voltage spikes in the switching-off process induced by L_{s1} in Figure 2.2 are modeled, which in fact could lead to the damage of the semiconductor switches. The diode is modeled as a short circuit when it is *on* and open circuit when it is *off*. No reverse recovery current of the diode is revealed. In reality, this current is included in the switching-on current of S_1. That is,

$$I_{on} = I + I_{rr} \qquad (2.3)$$

where I is the load current and I_{rr} is the reverse recovery current of diode D_1.

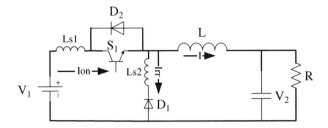

Figure 2.2   Buck topology considering short-timescale factors.

A current surge appears when S_1 turns on. As discussed in later chapters, these transients occur at the nanosecond level.

Another concern is the interaction between the ideal control algorithm and the actual implementation. Theoretically, the duty ratio D could vary from 0 to 1. However, the turn-on/off time of the semiconductor is not negligible. A minimum time interval is needed to effectively switch on/off the semiconductors, defined as the minimum pulse width (MPW) [6, 7]. Considering the minimum turn-on pulse width $T_{MPW\,on}$ and the minimum turn-off pulse width $T_{MPW\,off}$, technically

$$DT_s > T_{MPW\,on} \tag{2.4}$$

and

$$(1 - D)T_s > T_{MPW\,off} \tag{2.5}$$

Hence

$$D \in \left[\frac{T_{MPW\,on}}{T_s}, \, 1 - \frac{T_{MPW\,off}}{T_s} \right] \tag{2.6}$$

In other words, the duty ratio is restrained within a time interval determined by the MPW and switching period, which is different from the one derived from the ideal model of power electronic devices. MPW is a typical parameter varying from nanoseconds to microseconds depending on the types of power devices. The influence of MPW along with other typical indices and dead band will be presented in detail in Chapter 7.

2.2   Microelectronics vs. power electronics

Power electronics significantly differs from microelectronics. Microelectronics deals with the storage, transfer, and generation of information, or small signals, but power electronics aims at the storage, transfer, and generation of electrical energy. On the other hand, power electronics needs control circuits and logic to realize reliable and economic operation. It is therefore an integration of signals and energy. In modern power and energy systems, microelectronics is the brain of the system while power electronics is the muscle of the system.

Compared to microelectronics, research in power electronics is falling behind. For example, the computer science and communication fields have developed their own complete theoretical frameworks to direct the applications. Today, *power electronics* and *power electronic techniques* are in fact two different areas from the viewpoints of academia and engineering. However, at present, power electronics is hard to be defined as a *science* because the complete and accurate theoretical fundamentals are missing. In power electronics, circuit theory research, topology, and applications are the main targets of this field where linearized models are widely adopted. It should be pointed out that systems containing power semiconductors are far from "linearized models." In the last few decades, scholars and engineers have also paid lots of attention to non-linearized power electronic systems, such as chaos in DC–DC power supplies. However, up to now, in most of the power electronics applications, linear control theory, topologies and ideal semiconductor models are still used for the sake of research convenience, which presents a high risk in some circumstances or results in false conclusions. An understanding of power electronics remains on the stage of ideal, linearized models, and lumped parameters. Empirical coefficients and design principles are responsible for many problems in real-world applications.

2.2.1   Understanding semiconductor physics

Cost and reliability are the most important factors in power electronics design. Increasing the reliability is greatly associated with the selection of power electronic components. Generally, a qualified semiconductor switch should possess the following characteristics: high off-state blocking voltage, high current density, low on-state voltage drop, short switching-on/off intervals, high di/dt and dv/dt capability, and long life. The present controllable switches cannot offer the above requirements simultaneously. Hence, in practice, an engineer needs to consider the on-state, off-state, switching-on state, switching-off state, recovery, thermal, and mechanical characteristics all together and reach a reasonable trade off. Not only the steady-state performance, but also the transient processes should be studied.

Consider an IGBT as an example, as shown in Figure 2.3. The IGBT chips are connected by several thin strands of aluminum (wire bonding). Such a connection will result in stray parameters which contribute to high-voltage spikes under high di/dt. Meanwhile, from Figure 2.3b [8], the connection between the wire end and the bottom layer is too fragile to undertake high mechanical pressure, vibration, and harshness.

Thermal dissipation is another key issue to be addressed. Different operating temperatures result in different semiconductor endurance. For IGBTs, temperature rise is highly dependent on heat sink design and the electrical energy

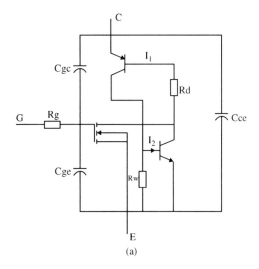

(a)

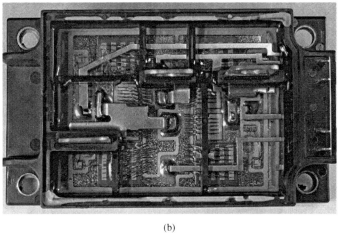

(b)

Figure 2.3   Inside configuration of an IGBT: (a) equivalent circuit of IGBT and (b) internal scheme of IGBT module.

generated inside the silicon chip. Figure 2.4 illustrates the maximum repetitive turn-off current versus chip temperature [9]. The higher the temperature, the lower the current ratings.

Thus a deep understanding of semiconductor physics is the premise of power electronics to offer safe and reliable application.

2.2.2   Evaluation of semiconductors

The reliability of power electronic systems is also determined by the interaction between different components, which results in the optimized

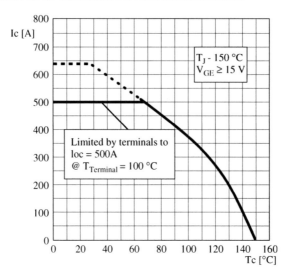

Figure 2.4   Current–temperature characteristics of an IGBT (SKM600GB1260). Courtesy Semikron.

design of topology, structure, and control strategies. The parameters shown in semiconductor datasheets are actually based on a specific single-switch test platform. They may vary in other systems with different structure, topology, stray parameters, and control algorithm. For example, a test setup of an IGCT is shown in Figure 2.5 [10].

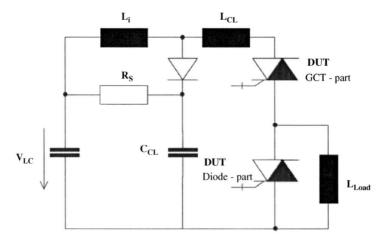

Figure 2.5   Test platform of an IGCT.

Due to variations in topology, manufacturing, and assembly technologies, some critical parameters such as stray inductance L_{CL} can hardly be guaranteed uniformity with the above test platform. The interaction between semiconductors and stray elements may distort the boundary of the safe operation area (SOA) [11] presented by the datasheet. In other words, the SOA based on a single-switch test platform is not always applicable to all operation modes and applications.

The problems listed above can be attributed to the characteristics of power electronics, as follows.

2.2.2.1   Multiple dimensions

Control systems, including microcontroller and gate-drive circuits, provide the control signals. The semiconductors, the main circuit, and the load constitute the energy flow paths and are tightly coupled with each other. Energy flow is not only a function of time, but also a function of space. Energy flows through the wire and at the same time, radiates into space in all directions. Power semiconductors, whose size is much larger than microelectronic switches, have a different manufacturing technology than microelectronics. Modern controlled semiconductors comprise a large number of series and parallel cells. For example, the diameter of the largest IGCT and GTO exceeds 100 mm and the cell count exceeds 100 000. The doping and implanting technologies cannot guarantee uniformity across all cells, which will lead to the diversity of the switch performance. For example, when a GTO turns on, some cells will turn on first while others remain off. The current will first flow through the cells turned on and create a localized heat rise.

For microelectronics, the above problems also exist but are not as severe. Small-sized IC components can reach uniformity easily. More importantly, the energy handled by the IC is negligible compared to power electronics.

2.2.2.2   High power

The voltage, current, and power capacity of power electronic semiconductors may exceed thousands of volts, amperes, and several megawatts, respectively. Experimental data shows that although the semiconductor consumes very little power in this process (it is mainly used to block or allow the energy to flow from source to load or load to source), the transient processes with high electromagnetic power could still damage the system. A single high-voltage spike lasting 1 μs will be enough to break down the switches.

2.2.2.3   Multimedia

In power electronic systems, semiconductors, bus bars, capacitors, inductors, resistors, cables, heat sinks, and loads are the media of power flow, while semiconductors constitute a path for power flow. It requires both the characteristics of microelectronics, such as slope, amplitude, and pulse width of gate-drive signals, and those of the power flows, such as voltage, current, dv/dt, and di/dt. At this point, not only should the pulse generation and load be taken into account, but also the media, information flow, and power flow carriers should be considered. For example, the traces from the gate-drive circuit to the gate should be modeled to evaluate the impact of the stray parameters along the wires. In order to suppress voltage spikes on semiconductors, the stray inductance needs to be minimized. This medium is no longer an "ideal conducting wire" as regarded in traditional research, but now becomes loads or energy storage components in the transient processes. Although their stored energy is very small compared to the overall system power rating, their power and energy dissipated in a small time interval could cause problems.

2.2.2.4   High nonlinearity and low predictability

The generation of microelectronic signals could be modeled and predicted effectively. However, the processing of these signals may encounter difficulties in power electronic systems. The pulses generated by the converter differ from those imposed on the load due to the existence of transmission cables. At the same time, other parameters, such as the dead band [12, 13], minimum pulse width, and variation of semiconductor characteristics shown in later chapters, make the real waveforms deviate from the expected ones. Some problems may occur unexpectedly, which could jeopardize system reliability and safety.

2.2.2.5   Different magnitudes of time constants

A power electronic converter contains power semiconductors, commutating loops, control modules, and load. On a higher level, it could be regarded as a combination of infotronics and electrical parts. A typical converter is shown in Figure 2.6.

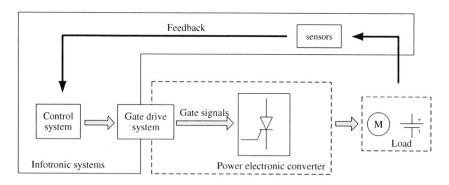

Figure 2.6   A typical power electronic converter.

In Figure 2.6, the time constants of energy conversion in each subsystem are different. For example, the time constants of commutating loops of passive devices are on the order of milliseconds. Those of power semiconductors and parts of control modules are on the order of microseconds, and the switching processes of semiconductors are in the range of nanoseconds. The converter could be connected with a mechanical load such as an electric motor, whose time constant ranges from milliseconds to seconds. The balance of energy generation, transmission, and storage is a key issue. Most of the failures in the subsystems and their components occur in the transients, that is, from one steady state to another. In these dynamic processes, an energy imbalance often arises, which leads to the local convergence of energy flow.

All the above characteristics can not be studied with the idealized models. A new research in the transient process is the inward search.

2.3   State of the art of research in short-timescale transients

Generally, the essence of all power electronic converter design is to implement the controllable and effective transfer of electromagnetic energy. In this process, the rules of energy conservation and continuity must be followed, which are the foundations of the investigation of transient processes. Traditional research on a macroscopic scale potentially leads to incomplete knowledge of such short-timescale transients in the systems. Some of these transients represent drastic and crucial energy transfer processes such as the switching-on/off transition of semiconductors. Hence, many modulation techniques are actually not feasible in real-world applications, especially in high-power converters where the design margin of power semiconductors is extremely limited.

Therefore, in order to achieve the reliability goals of the whole power electronic system, the short-timescale phenomena and their causes should be studied [14]. The energy flow and distribution in these processes should be carefully examined. The research perspective needs to switch from macroscopic to microscopic. Quantitative and mathematical methods need to be extensively used to study the transient energy flow. The ultimate goal is to combine traditional control algorithms and microprocesses to get high performance and high reliability simultaneously.

In terms of energy flow, the carriers of these short-timescale transients and their energy need to be characterized. Power electronic systems manipulate the energy flow through switching on and off the semiconductor devices. In this process, pulses are imposed on the gate and are regarded as control signals, directly corresponding to the power pulses on the main circuit and load. From an energy point of view, the pulse itself and the pulse sequence emerge as the carriers. Research on pulses and pulse sequences is the main line of this book.

2.3.1 Pulse definition

Robert G. Middleton defined a pulse in his book *Pulse Circuit Technology*, regarding it as an instantaneous impact of energy. One pulse could be considered as the summation of two step functions, one positive, going from zero to full amplitude, the other negative, going from zero to minus full amplitude. The pulse width is defined as the interval between the rising edge (RE) and trailing edge (TE). If RE leads TE, then the pulse is defined as positive, otherwise it is defined as negative [15].

Besides the basic pulse definition, there are other important properties of a pulse, as shown in Figure 2.7:

1. **Amplitude:** steady state value of the pulse.

2. **Rise time t_r:** time interval for the pulse to rise from 10% to 90% of its amplitude.

3. **Fall time t_f:** time interval for the pulse to fall from 90% to 10% of its amplitude.

4. **Pulse width t_w:** the time interval between RE and TE.

5. **Over-modulation σ:** σ_1, peak value minus steady-state amplitude, stands for the oscillation in RE; σ_2 symbolizes the oscillation in TE.

6. **Over-modulation time:** time intervals from a steady-state value to maximum then back to steady-state again.

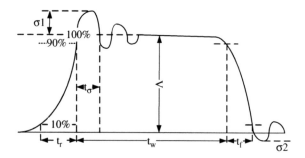

Figure 2.7   Definition of one pulse.

Pulse sequence, a combination of different pulses, reveal its own properties in addition to those of a single pulse:

1. **Time property:** if the pulses are periodic, then the pulse sequence can be described with a pulse repetition frequency (*prf* for simplicity). If the pulse sequence is periodic while inner pulses are not, then a pulse sequence repetition frequency (*psrf*) is used. If none of these apply, a pulse repetition rate (*prr*) is given to describe the pulse sequence mathematically.

Example 2.1   In the start-up process of a 380 V, 160 kW AC adjustable speed motor with DC pre-excitation [16], the output voltage and current are illustrated in Figure 2.8.

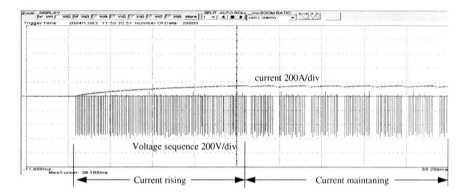

Figure 2.8   The waveforms of output voltage and current in DC pre-excitation.

Here two separate instances are illustrated. The first instance is the current rising process, in which the current is approaching the final value and only periodic voltage pulses are imposed on the motor terminals. The time interval between pulses is 250 µs, and all pulses have the same shape, hence they are periodic with $prf = 1/(250\,\mu s) = 4\,kHz$. The second instance is the current holding process, in which the control algorithm generates different voltage pulses so as to hold the current within a narrow hysteresis band. The periodic characteristic of a single pulse is replaced by that of a pulse sequence, with $psrf = 133\,Hz$ and $prr = 950\,Hz$. In this example, we can see that the RE, TE, width of a single pulse and the prf and prr of a pulse sequence are all important factors.

2. **Pulse relations:** consider the pulse sequence in Figure 2.9, where t_n is the start time of a given pulse. It represents the relative positions of different pulses. θ_n is the interval between the TE of the nth pulse and the RE of

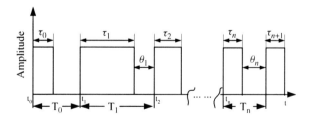

Figure 2.9   Pulse relations in one pulse sequence.

the $(n + 1)$th pulse. If $\theta_n > 0$, there is no overlap between the nth and $(n + 1)$th pulse. If $\theta_n < 0$, overlaps emerge, as in the case of the line-to-line voltage waveforms of multilevel inverters.

2.3.2   Pulsed energy and pulsed power

Pulses are functions of time and space and they always show continuity regardless of the traditional "0" or "1" representation of digital control. The time and space distribution of one pulse $u(x, t)$ can be expressed as

$$
\begin{cases}
u(x, t)_{|t=t_0^+} = u(x, t)_{|t=t_0^-} \\
u(x, t)_{|x=x_0^+} = u(x, t)_{|x=x_0^-}
\end{cases}
\tag{2.7}
$$

Pulses follow the transfer principles in time and space, shown as [17]

$$
k\frac{\partial^2 u}{\partial x^2} = \tau\frac{\partial^2 u}{\partial t^2} + \frac{\partial u}{\partial t}
\tag{2.8}
$$

where k and τ represent the properties of the media in which the pulse propagates. Equation 2.8 could have a different form for different pulses in different media. In power electronic systems, heat pulses and electromagnetic pulses can be described by Equation 2.8.

Pulses are also carriers of energy, which is determined by

$$
W(x) = \int_{t=0}^{+\infty} u(x, t)^2 dt
$$
$$
P(x, t) = u(x, t)^2
\tag{2.9}
$$

where $W(x)$ is the total energy of the pulse sequence. $P(x,t)$ can simply be regarded as power. In the short-timescale transient process, the energy carried by the pulse can be neglected, but the power of repeating pulses cannot be ignored. For the electromagnetic pulse, pulsed power is represented as [4]

$$
P = \oiint (E \times H)dS = -\oiiint \left(\frac{\partial\left(\varepsilon E^2/2 + uH^2/2\right)}{\partial t} + J \cdot E\right)dV
\tag{2.10}
$$

where P is power, E is the electric field density (V/m), H is the magnetic field density (A/m), J is the current density (A/m^2), ε is the dielectric coefficient (F/m), and μ is the permeability (H/m). Furthermore, Equation 2.10 can be expanded to

$$
\oiint (E \times H)dS = -\oiiint \left(\varepsilon E\frac{\partial E}{\partial t} + \mu H\frac{\partial H}{\partial t} + J \cdot E\right)dV
\tag{2.11}
$$

The first term on the right of Equation 2.11 is the energy contained in the electric field. The second term is the energy in the magnetic field and the third is the ohmic loss resulting in heat dissipation.

Example 2.2 The turn-off process of an IGCT in the circuit shown in Figure 2.10a is depicted in Figure 2.10b. Traditional calculations use the integration of voltage multiplied by current to estimate the switching loss, which is actually the third term in Equation 2.11. The first and second items are not considered.

The stages in this process are:

Stage 1 [t_0, t_1]: current remains constant while voltage starts to rise, increasing the electric field density. Hence the second and third items of Equation 2.11 are increasing.

Stage 2 [t_1, t_2]: voltage increases significantly while current starts to decrease. The magnetic energy decreases while electrical energy increases. The magnetic energy stored in the stray inductance is transformed to electrical energy in the IGCT.

Stage 3 [t_2, t_3]: the voltage slope remains fairly constant while current drops to zero abruptly. A voltage spike is induced by the stray inductance while the magnetic energy in the IGCT is neutralized.

Stage 4 [t_3, ∞]: voltage and current drop to a steady state, hence electrical and magnetic energy returns to a steady state.

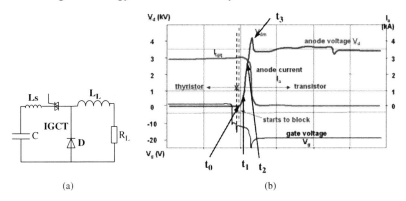

Figure 2.10   Experimental test of an IGCT: (a) buck circuit for one IGCT and (b) switching-off waveforms of one IGCT.

In the off state, voltage and current remain unchanged, making the first and second items of Equation 2.11 zero.

In this case, the dominant stage of energy transformation is stage 2. Furthermore, the first and second items of Equation 2.11 correspond to dv/dt and di/dt respectively. Therefore the calculation of losses should include the process of energy transformation, which cannot be revealed by integration of the product of voltage and current.

At a macroscopic level, Ferreira and van Wyk [5] divided the power electronic system into energy loops, energy modulation, and energy storage as shown in Figure 2.11a. As discussed above, the carrier of energy, pulse, and pulse sequence behave differently in power electronic systems. Each pulse behavior and its sequence correspond to a change in energy. Therefore, at a microscopic level, the research methodology shown in Figure 2.11a could be

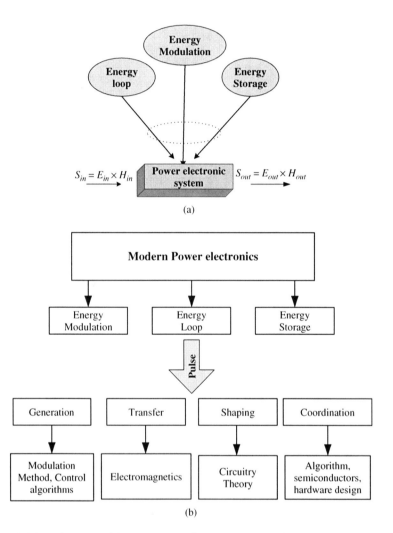

Figure 2.11   Theoretical framework and research methodology of modern power electronics: (a) theoretical framework targeting at energy and (b) research methodology aiming at pulse and pulse sequence.

refined to Figure 2.11b, categorizing the pulse behaviors as generation, transfer, shaping, and coordination.

At the present time, research on energy modulation in the power electronics field is prevalent. Control algorithms are well developed for different applications with various simulations methods. Research on energy loops includes: extraction of stray parameters and evaluation of their influence [18], selection of electro-magnetic materials [19], modeling of the transient energy propagation [20], and architecture design of electrical components [21]. Van Wyk *et al.* [22] analyzed the high-frequency skin-effect phenomenon, studied the design of copper traces and the bus bar, and precisely predicted the skin-effect location and depth, where a larger resistance material is used to mitigate the skin effect. This research is instructive in the design of simple-structure bus bars, especially in high-frequency and high-power-density converters. Rosado *et al.* [23] proposed a concept of "energy interface", which are key factors in the energy loop and partition the energy channels into subsections. The energy complies with conservation prin-ciples but behaves in different forms on the two sides of this interface. One paradigm is point A in Figure 2.12b. On the left side of A, the voltage and cur-rent are both DC components, while on the right side they are all AC components. This type of interface is widely existing in the power electronic building block (PEBB), as shown in Figure 2.12. In the transient processes, those interfaces need be paid more attention due to the potential energy convergence.

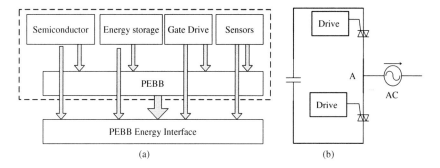

(a)   (b)

Figure 2.12   Energy interface in power electronic system: (a) energy interface [23] and (b) energy interface in one bridge.

It can be seen that all the above analyses no longer regard the energy loop as the energy transmission tunnel. Stray parameters on the energy loop behave as energy storage components in the commutating process, which significantly affect system reliability. From a microscopic perspective, the faster the semiconductors turn off, the more significant the transient power will be, and the more severe the electrical stress undertaken by the devices. These parameters include not only stray parameters, but also those parasitic parameters inside the semiconductors.

Following this perspective, research on energy loops is tightly coupled with energy storage. For energy storage, this research includes the modeling and simulation of advanced semiconductor switches and their commutating process, parameter selection of passive components, and electromagnetic integration [24]. Bhattacharyya and Sarnot [25] depicted the influence of junction capacitance in the switching-off process, which results in the displacement current in the semiconductor and potentially retriggers the device. In this book, only those storage parameters such as stray inductance in the energy loop will be modeled and investigated.

Modeling these energy transmission media from a system perspective offers realistic advantages. Caponet *et al.* [26] proposed a low-inductance multi-layer bus bar to attenuate voltage spikes on the semiconductors and reduce electromagnetic interference (EMI) in the turn-off process, which is now widely used in power electronic converters. Additionally, this design is simple and effective to avoid a complicated numerical calculation. A low-inductance bus bar is shown in Figure 2.13.

Figure 2.13   A low-inductance bus bar.

The difference in current, voltage, and electromagnetic field density at different places of the system determines the spatial arrangement of the core components. Therefore, modeling the energy loop directly provides the criteria of electromagnetic integration. Hull and Christopoulos [27] proposed that all power electronic devices could be modeled by lossless transmission line models, including loops, devices, and heat dissipation. In [28], the time–space distribution of the electromagnetic energy in a buck converter is calculated with lossless transmission line models. This model first detects the voltage or current waveform at a specific test point and then estimates the waveforms at any other point using Maxwell's equations.

2.4   Typical influential factors and transient processes

The research of transients in power electronic converters are associated with several aspects as described in the follows.

2.4.1   Failure mechanisms

There are various failure mechanisms, including overvoltage, overcurrent, turn-off voltage spike, and turn-on in-rush current of semiconductors (microsecond or nanosecond), the reverse recovery process of the freewheeling diode (nanosecond), and so on.

The fundamentals of power semiconductor switches lie on the PN junctions of semiconductors, whose main carriers are electrons and holes. Diffusion and drift occur in the semiconductors following the diffusion theory and Ohm's law, respectively. During normal operations, these two transportation methods maintain a dynamic balance. When operational conditions are not steady, this balance will be interrupted. The abrupt change in the distribution and transportation of carriers generates voltage spikes, overcurrent, current convergence, latch-up effects, local overheating, and so on. This will cause damage to the components and systems. For example, an IGCT in turn-on and turn-off processes behaves as a thyristor and transistor, respectively. This transformation is complete within $1\,\mu s$ and the physical characteristics are quite divergent. Some experimental waveforms of the voltage and current of an IGCT are illustrated in Figure 2.14.

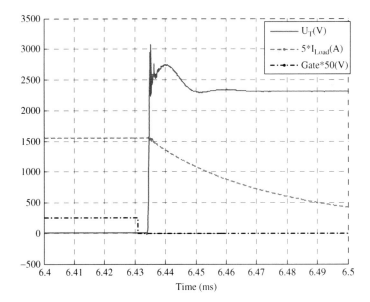

Figure 2.14   The turn-off voltage–current waveforms of an IGCT.

The significant overcurrent will induce a large switching loss, and current convergence, thus damaging the semiconductor. Figure 2.15 shows the surface of an IGCT destroyed with a partial burn and surface rupture, which confirms that power is not only a function of time but also a function of space. Different failure mechanisms lead to different damage in different locations. Research on failure mechanisms of GTOs shows, if the failure happens due to the high turn-off loss, the position of damage occurs in the middle of the silicon cells. If it is due to high di/dt in the turn-on process, the damage is on the edge. If it is due to the long-term overcurrent, the silicon cells burn out over large areas. Similar phenomena are also found in IGCTs (Figure 2.15a–d).

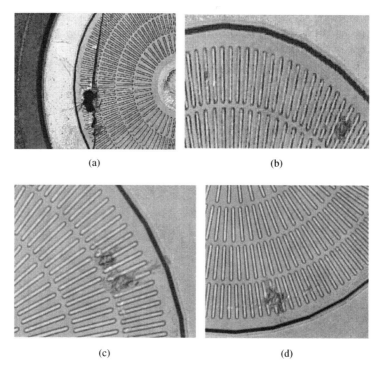

(a)   (b)

(c)   (d)

Figure 2.15   The section of destroyed IGCT in high-power inverter: (a) failure 1 (rupture in large area), (b) failure 2 (in the middle of silicon cell), (c) failure 3 (at the edge of silicon cell), and (d) failure 4 (across several silicon cells).

Various failure mechanisms exist in different semiconductors and applications. Only on the basis of the characteristics and internal structure of the device can an in-depth research be carried out on the failure mechanisms.

Example 2.3   One round chip of semiconductor turns off, as shown in Figure 2.16.

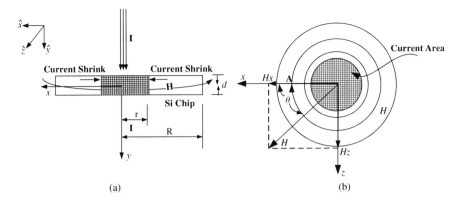

Figure 2.16   The current distribution of a switch in the turn-off process. (a) Current shrinking along the x-axis and (b) Current distributed in some area.

The diameter of the semiconductor is R. The gate is distributed on the edge. The gate signal pulse needs time to propagate from the edge to the center. Thus the edge cells turn off before the center cells. Current shrinks from the edge to the center. If the trailing time of current is t_f and the current is $i(t)$, according to Ampere's law,

$$H(x, t) = \begin{cases} \dfrac{i(t)}{2\pi r^2}x & x < r \\[2mm] \dfrac{i(t)}{2\pi x} & r < x < R \end{cases} \qquad (2.12)$$

Assuming $di(t)/dt = -I/t_f$, $i(0) = I$, then

$$\frac{\partial H_z}{\partial t} = \begin{cases} \dfrac{I}{2\pi r^2}\dfrac{x}{t_f} & x < r \\[2mm] \dfrac{I}{2\pi x t_f} & r < x < R \end{cases} \qquad (2.13)$$

$$H_x \equiv 0$$

According to Maxwell's equations,

$$\frac{\partial E_y}{\partial x} = -\left(\mu\frac{\partial H_z}{\partial t} + \mu\frac{\partial H_x}{\partial t}\right) = \begin{cases} -\dfrac{\mu I}{2\pi r^2}\dfrac{x}{t_f} & 0 < x < r \\[2mm] -\dfrac{\mu I}{2\pi x t_f} & r < x < R \end{cases} \qquad (2.14)$$

The boundary condition is $E_y(x = R) = E_R = U_0/d$, therefore

$$
E_y(x) = \int_R^x \frac{\partial E_y}{\partial x} dx
$$

$$
= \begin{cases} \dfrac{\mu I}{2\pi t_f}(\ln R - \ln r) + \dfrac{\mu I}{4\pi t_f}(r^2 - x^2) + E_R & x < r \\[4mm] \dfrac{\mu I}{2\pi t_f}(\ln R - \ln x) + E_R & r < x < R \end{cases} \tag{2.15}
$$

Hence the different points on the chip undertake different turn-off voltages. At $x = r$,

$$
V_y = E_y \cdot d
$$

$$
= \frac{\mu I}{2\pi t_f} d(\ln R - \ln r) + U_0 \tag{2.16}
$$

Example 2.3 shows that the semiconductor itself undertakes different voltage stress at different points. The first term of Equation 2.16 is the induced voltage associated with the dimensions of the semiconductor chip while the second term is the steady state voltage. It is worthwhile to point out that even with no stray inductance, a voltage spike will still exist in the switching-off process. The larger the current, the shorter the trailing-off time, and the higher the switch dimension, the higher the induced spike.

Essentially the energy transformation in a power electronic system is electromagnetic wave propagation, as indicated in [4]. This transformation is drastic in the transient process. Therefore, attention needs to be paid to the RE/TE of pulses to probe abnormal phenomena. If these problems root from the peripheral circuits, such as huge di/dt or heavy load the power electronics designer should select the right switches and set appropriate protection thresholds. In addition, the designer also needs to assign the necessary sensors and implement the proper protection to guarantee the converter to operate within its SOA.

2.4.2   Different parts of the main circuit

The structure of a power converter consists of the commutating loop, heat sink, and carrier loop, which are different channels of power flow. From an energy point of view, they are associated with different equations and time constants. The time constant of heat dissipation is in minutes, while that of a commutating loop is at the microsecond level and a semiconductor is at the nanosecond or even picosecond level. They interact with each other and maintain an energy balance at steady state. Different time constants and different speeds of energy flow make transients quite complicated with the impact of the stray parameters. Figure 2.17 shows three main systems in a power inverter.

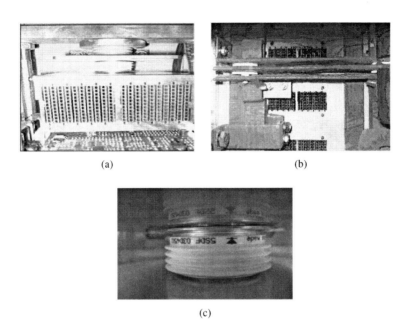

(a)   (b)

(c)

Figure 2.17   Three main parts in the three-level high-power inverter: (a) heat sink (minutes or hours), (b) DC bus (microseconds), and (c) carrier loop in semi-conductors (nanoseconds to picoseconds).

Although the time constants of different loops vary significantly, their energy flow is expressed by similar equations. The principle of heat transfer is

$$a\frac{\partial^2 T}{\partial x^2} = \tau_1 \frac{\partial^2 T}{\partial t^2} + \frac{\partial T}{\partial t} \tag{2.17}$$

The dynamic equations of voltage and current in the commutating loops are

$$k\frac{\partial^2 u}{\partial x^2} = \tau_2 \frac{\partial^2 u}{\partial t^2} + \frac{\partial u}{\partial t} \quad \text{and} \quad k\frac{\partial^2 i}{\partial x^2} = \tau_2 \frac{\partial^2 i}{\partial t^2} + \frac{\partial i}{\partial t} \tag{2.18}$$

The variables of T, u, and i are temperature, voltage, and current, respectively. k is the diffusion coefficient and τ_1 and τ_2 are time constants. From Equations 2.17 and 2.18, heat and electrical parameters have similar transportation theories with the exception that they have different coefficients. The analytical solution of Equation 2.18 is expressed as

$$u(x, t) = \begin{cases} 0 & t < x/c \\ u(0, t - x/c)e^{-(x/c)/2\tau_3} + \Delta U(x, t) & t > x/c \end{cases} \tag{2.19}$$

Here $c = \sqrt{\tau_3/k}$, and

$$\Delta U(x, t) = \frac{x}{2\sqrt{k\tau_3}} \int_{x/c}^{t} \left[I_1\left(\frac{1}{2\tau_3}\sqrt{\sigma^2 - \frac{x^2}{c^2}} \right) \Big/ \sqrt{\sigma^2 - \frac{x^2}{c^2}} \right]$$

$$e^{-\sigma/2\tau_3}u(0, t - \sigma)d\sigma$$

where I_1 is a first-order Bessel function,

$$I_1(x) = \sum_{k=0}^{\infty} \frac{1}{k!(k+1)!} \left(\frac{x}{2}\right)^{1+2k} \tag{2.20}$$

The function $i(x, t)$ is similar to Equation 2.19. The imbalance of energy distribution in the electrical process described by Equation 2.18 would lead to non-uniform heat distribution in space shown in Equation 2.17, and the local overheating would derate, degrade, or even damage the power semiconductor. Some practice in the assembly of power converters, such as long interconnects, large stray inductance in DC bus bars and a long distance between snubber circuits and power semiconductors, would impose extra electrical stress on the switches.

The coefficients in Equations 2.17 and 2.18 possess a strong nonlinearity in most cases, such as different characteristics of silicon devices under high and low voltages. Neglecting any of these propagations would pose a risk to the power electronic converter. Figure 2.18 shows the actual line-to-line voltage on the terminals of an electric motor, in comparison to the desired pulse. It can be seen that ripples exist at the pulse edges, which are caused by the transmission lines in the circuit. The commutating process cause differences in current, voltage, electromagnetic field, and a temperature distribution at different time and space. This directly determines the arrangement of key components in the converter. The definition and extraction of nonlinear parameters are of importance to analyze energy transportation and transformation in this system, as shown in later chapters.

2.4.3   Control modules and power system interacting with each other

The transients of a control system are in microseconds or nanoseconds. The digital control of systems in modern power electronics is mainly based on digital

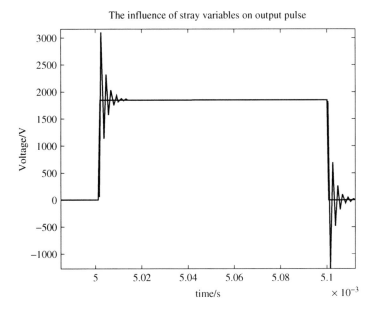

Figure 2.18   The line-to-line voltage on the motor terminals.

signal processing (DSP) and other multi-CPU networks, whose instruction execution time is on the order of microsecond or nanoseconds. In a real application, control systems and main power interact with each other. This interaction results in that macroscopic control strategies are not well implemented under some circumstances, which will be discussed in Chapter 7 and 8.

2.5   Methods to study the short-timescale transients

In order to investigate the short-timescale transients in power electronics, the following tasks need to be performed:

1. Use high-precision measurement, the precondition to detect transients effectively and accurately.

2. Model the semiconductor switches appropriately. Most of the short-timescale transients are caused by the switching processes. Therefore, ideal models of such switches are not helpful. However, overly complicated models such as semiconductor physics-based models use too many computational resources. A pure analytical model is difficult to establish for power electronic switches. Functional models outlining the electrical performance of the switch are a good choice under specific operation modes, but may not show good correlation with actual experimental

waveforms when the test conditions vary. A hybrid model combining the functional and analytical models could mitigate the above problems. Such models will be discussed in Chapter 3.

3. Model the peripheral circuits and components. For regular-shaped conductance lines, applying Maxwell's equations could yield accurate results. For other complex objects, finite element methods (FEMs) are a possible way to obtain stray parameters of the peripheral circuits, such as the bus bar, cable, and connection wires. Parasitic parameters of the passive components need to be addressed and determined.

4. Predict the energy flow path at any arbitrary time and location. The traditional averaged model is based on one switching period and not on REs/TEs of the pulse. Such a large-timescale model is not adequate for analyzing the microsecond and nanosecond energy transients, especially when most system failures occur in these transients. The energy flow in the system should be analyzed to find a potential location and time where energy tends to converge. In this process, traditional topology research should be used and further developed.

5. Implement macroscopic control algorithms at the microscopic level. On one hand, effective control strategies should be adopted on power electronic systems to realize reliable and efficient operation while also dealing with short-timescale phenomena, such as the dead band, minimum pulse width, and modulated error, as described in Chapters 7 and 8. On the other hand, the system should be optimized to take the most advantage of the control strategies, such as an EMI reactor to attenuate electromagnetic impulses and good placement of electrical components.

2.6 Summary

Research on short-timescale transients in power electronic systems is a challenging but meaningful task. The transients challenge the current theory of power electronics, which is based on averaged, idealized, or linearized models. It is possible that research in and application of nanosecond processes in power electronics will make some significant breakthroughs:

1. Theoretical innovations on short-timescale power flow. The viewpoint of power and energy would shift from the original *power = voltage × current* to electromagnetic based theory, to get rid of the ideal and lumped models. Nonlinear models should be established to help deepen the understanding of various failure mechanisms.

2. Theoretical innovations on traveling wave propagation. The transportation theory of inner carriers in semiconductors and free electrons in transmission lines could be comprehended. Thus the energy function is

extended to four-dimensional time and space functions. The energy flow and distribution at any time and any place will be estimated precisely.

3. Arbitrary waveform transformation. The ultimate goal of power electronics is to realize arbitrary waveform transformation. This transformation is not at the signal level, but at the energy level. In traditional fixed frequency electrical transmission, transformers are used in large-scale and complex power systems. In the DC and variable frequency AC area, a high-power and arbitrary wave transforming is needed. This form of energy transformation will benefit the future applications.

Based on the theory of power pulse phenomena, the traditional control strategies, component selection, device design, cooling calculation, and so on, could be interpreted into mathematical optimization problems and physical energy-level issues. The design of power electronic converters will be described and visualized in a more scientific way.

Research on power pulses in high-power inverters is an emerging field. The traditional pulsed power applications is mainly the single pulse which is generated by large capacitors and their combination and is also not fully controllable. The pulse sequence in power electronics is essentially different from pulsed power phenomena. The pulses in power electronics are sequential, repetitive, controllable, and generated by semiconductors based on the specific topology. This would find wider application as soon as power switches reach a breakthrough.

Research on transients in power electronic devices is beneficial to achieve arbitrary waveform transformation with high effectiveness and high reliability, resulting in precise analysis and effective simulation of the failure mechanisms of power switches. It is also crucial for estimating the switching characteristics, including calculations of switching losses and heat generation, arrangement of power commutating loops and peripheral circuits, quantitative analysis of EMI, and implementation of fault protection.

Finally, scientific research and applications cannot be isolated from each other. The research on transients in power electronics helps to establish the nonlinear model of power electronic switches and reflect device status under different operating conditions. By studying the electromagnetic energy distribution in high-power inverters, we are able to optimize the system, design effective auxiliary circuits, suppress dv/dt and di/dt, and control the transients of electromagnetic energy. The PWM algorithms could be further enhanced to realize an optimal combination of switching frequency, minimum pulse width, dead band, and modulation index with the fully understanding of short-timescale phenomena in power electronic systems.

References

1. Shen, J.Z., Xiong, Y., Cheng, X. *et al.* (2006) Power MOSFET switching loss analysis: a new insight. Industry Applications Conference, pp. 1438–1442.

2. Rodríguez, M., Rodríguez, A., Miaja, P.F. *et al.* (2010) An insight into the switching process of power MOSFETs: an improved analytical losses model. *IEEE Transactions on Power Electronics*, **25** (6), 1626–1640.

3. Jovanovic, M.M. (1997) A technique for reducing rectifier reverse-recovery-related losses in high-voltage, high-power boost converters. Applied Power Electronics Conference and Exposition, pp. 1000–1007.

4. Yamazaki, M., Kobayashi, H., and Shinohara, S. (2004) Forward transient behavior of PiN and super junction diodes. Power Semiconductor Devices and ICs, pp. 197–200.

5. Ferreira, J.A.N. and van Wyk, J.D. (2001) Electromagnetic energy propagation in power electronic converters toward future electromagnetic integration. *Proceedings of the IEEE*, **86** (6), 876–889.

6. Welchko, B.A., Schulz, S.E., and Hiti, S. (2006) Effects and compensation of dead-time and minimum pulse-width limitations in two-level PWM voltage source inverters. Industry Applications Conference, pp. 889–896.

7. Bhattacharyya, B.K., Levin, A., and Gang, H. (2007) A semi-empirical approach to determine the effective minimum current pulse width for an operating silicon chip. Power Engineering Conference, pp. 922–927.

8. Daniel, D. (2007) L'intérieur d'un module (bras de pont) IGBT 400 A/600 V, technologie PT, incluant 2 x 2 IGBT et 2 x 2 diodes ainsi que les circuits de protection de gate et de surintensité. http://en.wikipedia.org/wiki/File:IGBT_2441.JPG (accessed March 23, 2011).

9. SEMIKRON (2005) IGBT Modules SKM 453A020, SEMIKRON, Germany, 05-04-2005 SEN.

10. ABB Semiconductors (2003) Reverse Conducting Integrated Gate-Commutated Thyristor 5SHX 08F4502, ABB Switzerland Ltd.

11. Yilmaz, H., Tsui, T., Bencuya, I. *et al.* (1989) Safe operating area of power DMOS-FETs. Power Electronics Specialists Conference, pp. 170–175.

12. Liu, Yu., Huang, A.Q., and Bhattacharya, S. (2009) Dead-band controller for balancing individual DC capacitor voltages in cascade multilevel inverter based STATCOM. Applied Power Electronics Conference and Exposition, pp. 683–688.

13. Hosseini, S.H. and Sabahi, M. (2006) Three-phase active filter using a single-phase STATCOM structure with asymmetrical dead-band control. Power Electronics and Motion Control Conference, pp. 1–6.

14. Zhengming, Z., Hua, B., and Liqiang, Y. (2007) Transient of power pulse and its sequence in power electronics. *Science in China Series E: Technological Sciences*, **50** (3), 351–360.

15. Middleton, R.G. (1964) *Pulse Circuit Technology*, Howard W. Sams, Indianapolis.

16. Hua, B., Zhengming, Z., Liqiang, Y., and Bing, L. (2006) A high voltage and high power adjustable speed drive system using the integrated LC and step-up transforming filter. *IEEE Transactions on Power Electronics*, **21** (5), 1336–1346.

17. Cabeceira, A.C.L., Grande, A., Barba, I., and Represa, J. (2004) A 2D-TLM model for electromagnetic wave propagation in chiral media. Antennas and Propagation Society International Symposium, pp. 1487–1490.

18. Fasching, M. (1996) Effects of stray inductance on switching transients of an IGBT propulsion inverter and resulting gate control strategies. AFRICON, pp. 467–472.

19. Van Wyk, J.D. Jr., Cronje, W.A., van Wyk, J.D. *et al.* (2005) Power electronic interconnects: skin- and proximity-effect-based frequency selective multi-path propagation. *IEEE Transactions on Power Electronics*, **20** (3), 600–610.

20. Bai, H., Mi, C.C., and Gargies, S. (2008) The short-time-scale transient processes in high-voltage and high-power isolated bidirectional DC-DC converters. *IEEE Transactions on Power Electronics*, **23** (6), 2648–2656.

21. Shenai, K., Neudeck, P.G., and Schwarze, G. (2000) Design and technology of compact high-power converters. Energy Conversion Engineering Conference and Exhibit, pp. 30–36.

22. Van Wyk, J.D. Jr., Cronje, W.A., van Wyk, J.D. *et al.* (2005) Power electronic interconnects: skin- and proximity-effect-based frequency selective multi-path propagation. *IEEE Transactions on Power Electronics*, **20** (3), 600–610.

23. Rosado, S., Prasai, A., Wang, F., and Boroyevich, D. (2005) Study of the energy flow characteristics in power electronic conversion systems. Electric Ship Technologies Symposium, pp. 333–339.

24. Yuan, L., Zhao, Z., Yi, R. *et al.* (2007) Performance evaluation of switch devices equipped in high-power three-level inverters. *IEEE Transactions on Industrial Electronics*, **54** (6), 2993–3000.

25. Bhattacharyya, A.B. and Sarnot, S.L. (1967) Effect of the junction capacitance variation and a small inductance on the switching time of a tunnel diode. *Proceedings of the IEEE*, **58** (12), 1957–1959.

26. Caponet, M.C., Profumo, F., and Tenconi, A. (2002) Low stray inductance bus bar design and construction for good EMC performance in power electronic circuit. *IEEE Transactions on Power Electronics*, **17** (2), 225–231.

27. Hui, S.Y.R. and Christopoulos, C. (1990) A discrete approach to the modeling of power electronic switching networks. *IEEE Transactions on Power Electronics*, **5** (4), 398–403.

28. Dongyan, W., Linchang, Z., and Kesheng, Z. (1997) Transient-field characterization of reverse recovery behavior of diode with interconnection. International Symposium on Electromagnetic Compatibility, pp. 630–633.

3

Power semiconductor devices, integrated power circuits, and their short-timescale transients

3.1 Major characteristics of semiconductors

Semiconductors are the building blocks of power electronic converters. The topology and architecture of power converters are driven by innovations in semiconductor devices. As discussed in Chapter 2, semiconductors have various electrical, mechanical, and thermal characteristics which need to be well thought out in the design and analysis of power converters. More specifically, the electrical characteristics include six major aspects: switching on, switching off, on state, off state, triggering, and recovery. Thermal and mechanical characteristics, such as vibration, fatigue, and stress, are also closely related to the electrical characteristics.

Switching-on and switching-off characteristics reveal the dynamic switching processes of semiconductor devices. Due to the overlap region of current and voltage, losses in switching processes are inevitable. Therefore, the switching loss has to be addressed in the design. In addition, the EMI caused by di/dt and dv/dt in these switching processes could affect operation of the system. Reduction of these losses and mitigation of EMI issues can be achieved by improving the control algorithms, such as soft-switching control depicted in later chapters.

On/off-state characteristics are mainly described by the on/off resistances, which directly correspond to the on-state voltage drop and off-state leakage current. To evaluate whether the semiconductors are suitable for some specific

Transients of Modern Power Electronics, First Edition. Hua Bai and Chris Mi.
© 2011 John Wiley & Sons, Ltd. Published 2011 by John Wiley & Sons, Ltd.

applications, the maximum breakdown voltage and maximum repetitive turn-off current are other important parameters to be considered in the design process.

Triggering characteristics include turn-on delay, turn-off delay, amplitude of triggering signals, and minimum triggering width. Sometimes a retriggering circuit is needed to ensure the device is fully turned on. Recovery characteristics include the forward recovery process (forward voltage) and reverse recovery process (recovery time, reverse current peak, and reverse voltage peak).

Thermal characteristics determine the highest temperature at which the switches can operate. Most semiconductor devices will degrade at high temperature [1]. In order to remove the excessive heat generated inside the chip, a cooling loop is needed.

Mechanical characteristics determine how to package and assemble the semiconductor switches. For example, the mechanical pressure on the switches should not exceed the upper limit that the switches are designed for, otherwise the packaging could be damaged. Vibration, fatigue, and stress can cause degradation of the device or faults during operation.

The items listed above will be discussed in detail for different semiconductors in the following sections. When a switch is selected, its characteristics should meet the requirements of the system. We need to know the feasibility and predict its actual behavior in various operating scenarios. The research methodology can include modeling, simulation, experiments, and so on. Modeling of the power devices is one of the most important steps in the study of short-timescale transient processes of power electronic converters.

3.2   Modeling methods of semiconductors

A comprehensive model of a semiconductor device should include all of the characteristics listed in Section 3.1. In addition, the semiconductor models have to interact with other components coexisting in the same circuit, and be able to adapt to different architectures and control algorithms. Present modeling methods include the physics-based model, functional model, average model, and hybrid model:

1. **Physics-based model:** these types of models are based on the physical mechanisms inside the semiconductor, trying to reveal the real behavior of the inside carriers. Kraus and Hoffmann [2] proposed a physics-based model of an IGBT and a diode, which reveals the switching-on, switching-off, and reverse recovery characteristics. Physics-based modeling requires a large amount of effort on the determination of parameters, such as the doping concentration, the section area of the chip, and diffusion depth. The results are accurate but the calculation speed is slow.

2. **Functional model:** the functional model of a semiconductor reveals the functions of the devices without going into detail of what happens inside the semiconductor. Prior to this modeling, tests should be pursued to get experimental data such as voltage and current limits. The model is established to fit well with experimental data. Yuan *et al.* [3] set up the functional model of an IGCT based on PSIM software and the model was effectively used to study the performance of series-connected IGCTs in high-power and high-voltage inverters [4]. Functional modeling expedites the simulation speed and facilitates the modeling procedure, but the precision cannot be guaranteed under different operating conditions.

3. **Average model:** an average model of a semiconductor device is developed to reveal the major external behaviors of the devices at steady state or in the transient state. Average models can take into account the nonlinearity, losses, and other main parameters, but do not contain the details of the switching process of the device.

4. **Hybrid model:** the physics-based model of semiconductors requires more detailed data inside the chip, which makes the modeling process difficult and slows the simulation speed. On the other hand, functional modeling does not reveal the physics of the devices. Therefore, combining the two different modeling approaches can make the modeling process more efficient without losing critical information.

3.2.1   Hybrid model of a diode

The most popular modeling method for the diode is the lumped-charge method [5]. The reverse recovery of the diode is illustrated in Figure 3.1a. The disadvantages of the lumped-charge model are: (i) although each modeled segment matches the known physical mechanism, at $t = T_2$, di/dt of the reverse recovery process becomes discontinuous; and (ii) the slope of the reverse recovery current does not follow the real process. In order to accurately describe the actual process, some modifications are necessary, as shown Figure 3.1b.

A hybrid model is then established to describe the reverse recovery process by four segments as shown in Figure 3.1b and Table 3.1 [6]. Some portion of the model is physics based and other portions are functional descriptions. In the table, the mathematical description of processes 1 and 3 is based on the physical process. Other processes correspond to the functional modeling. The relevant coefficients are determined by experimental data.

3.3   IGBT

The IGBT is a three-terminal power semiconductor device, notable for its high-current capability and fast switching. It is used in many applications, including

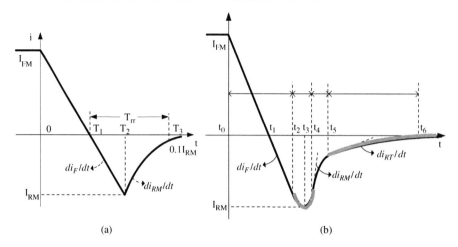

(a)   (b)

*Figure 3.1   Comparison of the lumped-charge model and the hybrid model:
(a) traditional lumped-charge model and (b) hybrid model.* © *[2008] Reprinted,
with permission, from Transactions of China Electrotechnical Society.*

Table 3.1   Hybrid model of the recovery process of diodes.

1	$i = I_{FM} - at \, (0 < t \le t_2)$	
2.1	$i = -(1 - k_1)I_{RM} \sin[(\omega_F(t - t_2)] \\ \quad - k_1 I_{RM} \quad (t_2 < t \le t_3)$	$di/dt \mid_{t=t_2} = -(1 - k_1)I_{RM}\omega_F \\ \qquad\qquad = di_F/dt$
2.2	$i_D = -(1 - k_1)I_{RM} \sin[\omega_R \, (t - t_3) \\ \quad + \pi/2] - k_1 I_{RM} \, (t_3 < t \le t_4)$	$di/dt \mid_{t=t_4} = (1 - k_1)I_{RM} \\ \quad \omega_R = di_{RM}/dt$
3	$i = -q_M/t = i(t_4)e^{-[(t-t_4)/\tau_{rr}]} \\ \quad (t_4 \le t \le t_5)$	$di_{RM}/dt = -k_1 I_{RM}/\tau_{rr}$
4	$i = i(t_5)e^{-[(t-t_5)/\tau_{rt}]} \\ \quad (t_5 \le t \le t_6)$	$di_{RT}/dt = -k_2 I_{RM}/\tau_{rt}$

variable speed motor drives, power system applications, switched-mode power
supplies, residential applications (variable speed refrigerator compressor, washer,
air-conditioner, and even audio systems with switching amplifiers), automobiles,
communications, induction heating, and many other areas. Large IGBT modules
typically consist of many devices in parallel and can have very high current
handling capabilities, on the order of hundreds of amperes with blocking voltages
up to 6000 V [7].

Figure 3.2 shows the structure of an IGBT. On one hand, the IGBT can
be regarded as a device combining the simple gate-drive characteristics of a

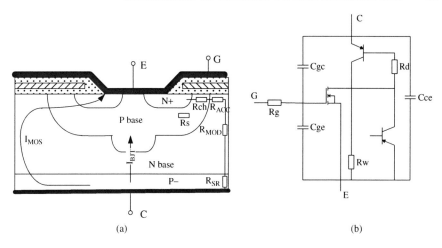

Figure 3.2   Functional model of IGBT: (a) cross-section of an IGBT and (b) functional model of an IGBT.

MOSFET with a high-current and low-saturation-voltage bipolar transistor. On other hand, we can also think of an IGBT as a MOSFET plus a P-substrate forming a transistor structure. The particulars of the IGBT determine its operating principle as follows:

1. **Conductance modulating effect [8]:**   all bipolar devices have this effect. The higher the current, the lower the internal resistance, opposite to the ohmic effect.

2. **Ohmic effect:** the internal MOSFET is a unipolar device. The higher the current, the higher the internal resistance.

3. **Current-clutching effect [9]:**   modern IGBT modules are made of many single cells. For the bipolar-transistor, due to the existence of the conductance modulating effect, the higher current will induce a lower resistance, therefore during the switching-on process, the current will tend to concentrate in some specific cells. Thus the current is not evenly distributed among the cells. Some cells will experience higher current during the switching-on process, therefore operating at a higher temperature.

4. **Latch-up effect [10]:** the IGBT has a parasitic thyristor. High current may prevent the device from turning off.

Figure 3.2b shows the functional model of an IGBT. The parameters are described as follows:

- C_{ge} – gate–emitter capacitance;

- C_{ce} – collector–emitter capacitance;

- C_{gc} – gate–collector capacitance;

- R_g – gate resistance;

- R_d – resistance of the N drift region, $R_{ACC} + R_{MOD}$ in Figure 3.2a;

- R_w – resistance of P drift region, Rs_1 in Figure 3.2a.

3.4 IGCT

GTOs (thyristors) were the primary choices for high-power applications before IGBTs and IGCTs were invented. However, the GTO has a complicated gate-drive circuit to deal with. The gate-commutated thyristor (GCT) is one of the newest power switches to take advantage of the GTO integrating more flexible and efficient gate control. A GCT is a semiconductor based on the GTO structure, whose cathode emitter can be turned off instantaneously like a transistor. It therefore has the low conduction loss of a thyristor as well as the low switching loss and high dv/dt capability of a bipolar transistor. Manufacturability and reliability are also improved. The IGCT is the combination of the GCT device and a low-inductance gate unit. This technology extends transistor performance to well above the megawatt range, with 4.5 kV devices capable of turning off currents up to 4 kA, and 6 kV devices capable of turning off currents up to 3 kA.

Structures of IGCTs are shown in Figure 3.3 [11]. At present, only one kind of IGCT is available, that is, the planar package.

Power semiconductor switches are based on PN junctions, whose main carriers are electrons and holes, whether the device is a diode, an IGBT, or an IGCT (Schottky diode is an exception). Carrier movement in the semiconductors follows the diffusion theory and Ohm's law. With external constraints, both diffusion and drift maintain a dynamic balance. When operating conditions are not stable, this balance is interrupted. The abrupt change in the distribution and transportation of carriers causes voltage spikes, overcurrent, current divergence, latch-up effect, local overheating, and so on. This would damage the components if this excessive energy is not dissipated in time. For example, the IGCT during turn-on and turn-off processes behaves as a thyristor and a transistor, respectively. Some experimental waveforms of the voltage and current on an IGCT are illustrated in Figure 3.4 [12].

The significant overcurrent would result in a large surface pressure, large switching-off loss, and current concentration, therefore potentially damaging the semiconductor. Figure 2.15 showed the surface of one IGCT destroyed with a partial burn and surface rupture. Here the power pulse is not only a function of time but also a function of space.

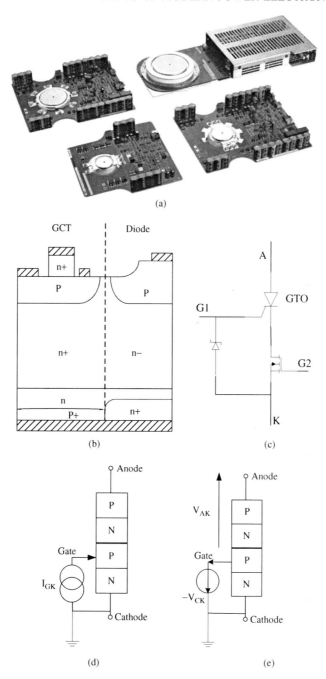

Figure 3.3   IGCT and its operational theory: (a) four types of IGCTs, (b) struc-ture of IGCT, (c) functional model of IGCTs, (d) turn-on structure, and (e) turn-off structure.

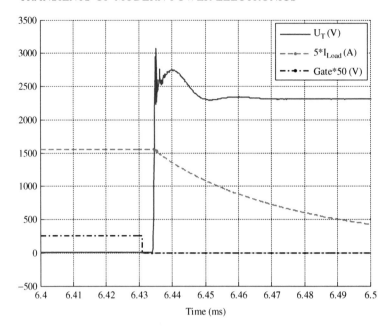

Figure 3.4   The turn-off voltage–current waveforms of an IGCT.

3.5   Silicon carbide junction field effect transistor*

Silicon carbide (SiC) is a new material that can be used to produce semiconductor devices. Various SiC-centered switches have been developed, such as the SiC MOSFET, SiC IGBT, and SiC JFET (Junction Field Effect Transistor) [13]. While the SiC MOSFET and SiC IGBT are difficult to fabricate at the present time, the SiC JFET is easier to manufacture (to the date of this book, Semisouth has already launched 1200V/30A SiC MOSFET.). Therefore, of all the SiC power transistors currently under development (e.g., MOSFETs, JFETs, BJTs, IGBTs), SiC JFETs have great potential for near-term real-world applications. With extensive research and development efforts under way, it is anticipated that more SiC JFETs will be used in power electronic systems.

The biggest barrier to using SiC JFETs is its "normally on" characteristic [14]. The threshold voltage to turn on SiC JFETs is less than zero volts. In other words, a negative gate voltage is necessary to turn off a SiC JFET. When a negative voltage is imposed on the gate, the channel developed by the electric field will be narrow with the increased channel resistance. When the negative voltage is sufficient, the device will be completely turned off. This characteristic is not desirable in real-world applications, for example, a shoot through of the DC bus could happen if the gate-drive voltage drops to zero in the SiC JFET based voltage source inverter.

* © [2010] Inderscience. Reprinted, with permission, from the International Journal of Electric and Hybrid Vehicles.

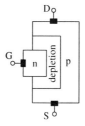

Figure 3.5   SiC JFET [15].

Many solutions have been proposed to effectively use SiC JFETs. One feasible solution is to use a SiC JFET (Figure 3.5) cascaded with a low on-resistance Si MOSFET to form a normally off SiC switch [16]. The control becomes easier but the additional Si MOSFET will degrade the JFET performance.

Another solution is to use a negative voltage to turn off the SiC JFET switches. A functional model of SiC JFETs is needed to aid the circuit design (the selection of gate-drive parameters), testing (transient electrical stress), prototyping, and product development (high-temperature operation). Some JFET models have been developed based on existing models of Si-based power devices [17]. However, these models have large deviations from the actual SiC devices in both static and dynamic performance.

Different approaches can be used to develop models of power electronic devices. The model of a SiC device developed in this chapter adopts the method proposed in [18]. In this model, a voltage-controlled current source is used to simulate the current in the JFET channel as shown in Equation 3.1. The junction capacitors are expressed in Equations 3.2 and 3.5. Since the main concern is the transient process, the on/off-state equations are neglected:

$$I_{JFET} = i_0 \left\{ 1 + \tanh \left[P(V_{gs} - V_{th}) \right] \right\} \cdot \tanh(\alpha V_{ds}) e^{\lambda V_{ds}} \quad \text{if } V_{ds} \geq 0 \qquad (3.1)$$

$$W_{gs} = \sqrt{\frac{2\varepsilon \left| V_{gs} - V_{th} \right|}{q N_b}} \qquad (3.2)$$

$$W_{gd} = \sqrt{\frac{2\varepsilon \left| V_{gd} - V_{th} \right|}{q N_b}} \qquad (3.3)$$

$$C_{gs} = \frac{f_{csj} A_s \varepsilon}{W_{gs}} \qquad (3.4)$$

$$C_{gd} = \frac{A_g \varepsilon}{W_{gd}} \qquad (3.5)$$

where V_{gs} is the gate voltage, V_{gd} is the voltage across the gate to the drain, V_{th} is the threshold voltage to turn on/off the JFET, and V_{ds} is the voltage across the drain to the source. W_{gs} is the depletion width of the gate–source area, W_{gd} is the depletion width of the gate–drain area, q is the fundamental electronic charge, N_b is the base dopant density, and A_g and A_s are the gate–drain and gate–source overlap areas, respectively.

In Equation 3.1, the first term represents the conductivity modulation by the gate voltage, and the second term describes the nonlinear relationship between V_{ds} and I_{ds}. In order to determine the parameters in Equations 3.1–3.5, the voltage–current waveforms under specific operating conditions are tested to fit the waveforms with simulated results, by adjusting the relevant parameters in the functional model of Equations 3.1–3.5 during simulation.

The SiC module used for the test is shown in Figure 3.6a. Two parallel-connected SiC JFETs are on the left and a SiC freewheeling diode is on the right which is a SiC Schottky diode used to eliminate the reverse recovery current. Because of the "normally on" feature, the SiC JFET turns on at zero gate voltage. A negative voltage, usually −24 to −30 V, is necessary to turn off the SiC JFET.

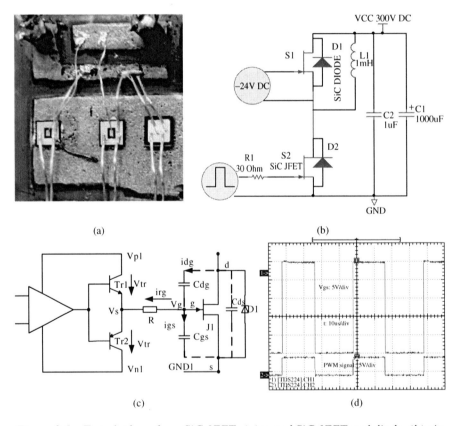

Figure 3.6   *Test platform for a SiC JFET: (a) tested SiC JFET and diode, (b) circuit configuration, (c) gate-drive circuit, and (d) gate-drive waveforms.*

In this design, an optical coupler is used to isolate the upper and lower driver circuit in the same bridge, as shown in Figure 3.6b. The totem pole stage amplifies the driver capability followed by the gate resistor as in shown Figure 3.6c. The gating control logic is shown in Figure 3.6d. When a PWM signal from the digital controller is high, the gating voltage is equal to V_{CC} (0–5 V) and the SiC JFET turns on. Otherwise, the gating voltage is equal to V_{SS} (−24 to −30 V) and the SiC JFET turns off.

The transient performance of the SiC JFET is simulated based on Equations 3.1–3.5. The turn-on and turn-off voltage–current waveforms are compared to the experimental results and the relevant parameters are adjusted to match the waveforms. Figure 3.7 shows a comparison of experimental and simulated switch-on/off processes of a SiC JFET at 300 V/8 A with a gate-drive resistor equal to 1 Ω. The simulated results coincide with experimental waveforms for both turn-on and turn-off processes.

SiC devices can handle high voltages up to 1200 V. However, present technologies determine they can hardly handle very large currents. Typical individual SiC JFET chips are rated around 10–15 A. Figure 3.8 shows a module ruined by overcurrent.

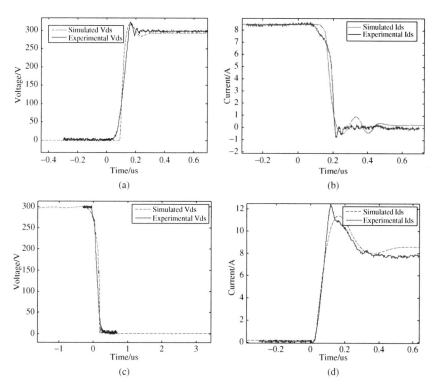

Figure 3.7   Comparison of simulated and experimental results: (a) turn-off voltage V_{ds}, (b) turn-off current I_{ds}, (c) turn-on voltage V_{ds}, and (d) turn-on current I_{ds}.

Figure 3.8   SiC JFET module destroyed by overcurrent.

3.6   System-level SOA

Semiconductor models, in particular the functional models, are helpful to analyze the operation and various failure mechanisms. In addition, datasheets of semiconductor devices provided from measurements on test circuits are helpful to set the operating limits of a given system with certain types of devices. However, the diversified power electronic circuit topologies determine that both functional models and datasheets of the devices will be different in different applications since some key parameters in a real system cannot be assumed to be the same as those in the test circuit, such as stray inductance. The diversity of load condition, operation modes, and peripheral circuits for the semiconductors determines that the SOA for a single-chip semiconductor is not always the same as the system-level safe operation area (SSOA). The SSOA is determined by the combination of energy loop, energy modulating strategy, and energy storage. Zhao *et al.* [19] explained the SSOA of a three-level inverter where the DC-link voltage, peak current, dv/dt, di/dt, and output power were included in the analysis to identify various failure modes. Mi *et al.* [20] elaborated on the SSOA through a tradeoff selection of leakage inductance to enhance the power capability and suppress maximum current peaks. The above research is beneficial for system reliability in the dynamic process under special topologies and circuits.

The determination of the SSOA is a complicated task, which not only needs to involve various possible macroscopic operational states, but also should comprehensively extract the different parameters of the system and evaluate their influence under the relevant control strategy. Essentially, the SSOA synthesizes the energy loop, energy storage, and energy modulation.

3.6.1   Case 1: System-level SOA of a three-level DC–AC inverter*

Figure 3.9 shows the topology of a typical three-level DC–AC inverter.

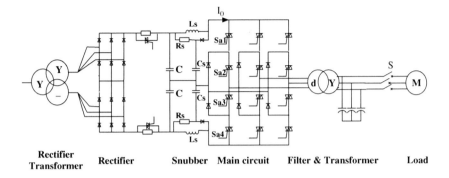

Figure 3.9   The main circuit topology of the three-level NPC inverter.

In this topology, each bridge is equipped with four semiconductors, for example, Sa1–Sa4 in the first bridge. When the topology is complicated, the interactions among the different components become important. For the IGCT-based, high-voltage, three-level neutral point clamped (NPC) inverter shown in Figure 3.9, the DC voltage is inversely proportional to the output current of the inverter when a constant output power is maintained. If the DC bus voltage is too high, it can damage the IGCT. But if the voltage is too low, it will result in high output current in order to maintain the rated output power. A high-voltage spike on the IGCT will be induced by the parasitic inductance during the turn-off process. High-voltage spikes can also damage the semiconductor device. The SSOA must ensure overall system safety as shown in Figure 3.10, where the horizontal axis is one-half of the DC-link voltage and the vertical axis represents the turn-off current. The light gray area is for a single IGCT defined by the datasheet, and the darker gray area is for the SSOA. The SSOA here considers the electrical endurance of the IGCT, real operating conditions, and stray parameters. The design of the system should not exceed the SSOA [19].

3.6.2   Case 2: System-level SOA of a bidirectional DC–DC converter**

Figure 3.11 shows the topology of a 10 kW, 600 V isolated bidirectional DC–DC converter based on a dual active bridge (DAB).

* © [2009] IEEE. Reprinted, with permission, from IEEE Transactions on Industrial Electronics.
** © [2008] IET. Reprinted, with permission, from IET Power Electronics.

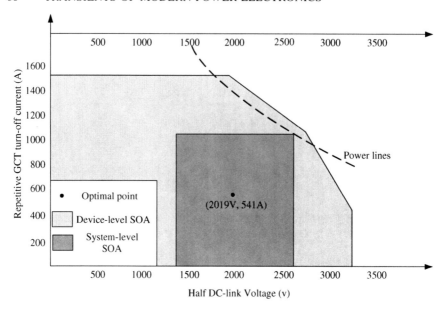

Figure 3.10   System-level SOA of an IGCT-based three-level inverter.

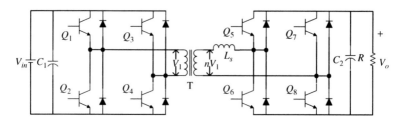

Figure 3.11   The isolated DAB bidirectional DC–DC converter system in a HEV.

In this converter, the voltage stress can be handled easily. But since the supply and the load can both be voltage sources, the current impact must be considered. The maximum current will be constrained by the leakage inductance L_s of the isolation transformer. However, a larger L_s will decrease the output power for the given device rating. Hence the key step for SSOA design of the bidirectional DC–DC converter is the selection of L_s and evaluation of its impact on current and output power, as shown in Figure 3.12. For example, if the switching frequency is chosen as 10 kHz, the leakage inductance should be chosen within the interval defined by A3 in Figure 3.12 [20].

3.6.3   Case 3: System-level SOA of an EV battery charger

The isolated DC–DC converter shown in Figure 3.13 takes a DC input and produces a DC output voltage. In order to take an AC input, an extra rectifier is

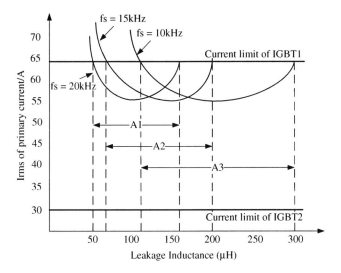

Figure 3.12   System-level SOA of a bidirectional DC–DC converter.

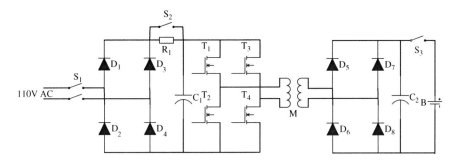

Figure 3.13   Schematic of the whole system.

added. The power factor correction stage is neglected in the circuit for simplicity of analysis. Here the DC-bus voltage on the primary side is V_1, the output voltage is V_2, the equivalent leakage inductance of the secondary side of the transformer is L_s, the turns ratio is n, and the switching frequency of the semiconductor switches is f_s. The input voltage of the system is 110 V AC from the electric grid. When S_1 is on and S_2 is off, the primary capacitor is charged through R_1 and the diode rectifier through D_1–D_4. After C_1 is charged to the rated voltage, S_2 will be turned on. A pre-charging stage is done. During this period, inverter bridges remain idle and S_3 stays off.

After detecting the start command with the aid of peripheral circuits, the PWM signals are generated to trigger MOSFETs T_1–T_4 with a phase-shift control to charge C_2 through the isolated transformer M and the secondary rectifier bridge

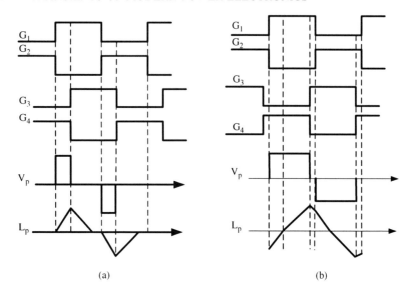

(a)   (b)

Figure 3.14   Control strategy of an isolated DC–DC converter: (a) in discontinuous mode and (b) in continuous mode.

D_5-D_8. The time sequence of the gate signals is shown in Figure 3.14a,b, where G_1-G_4 are the gate signals on T_1-T_4. Therefore, in discontinuous mode, T_1 and T_2 are in zero current switching (ZCS) mode. This topology can easily realize soft switching.

When the voltage of C_2 reaches the battery voltage V_b, S_3 is turned on. Then the system enters charging mode operation.

In the discontinuous mode, as shown Figure 3.14,

$$I_{\max} = \frac{nD}{2L_s f_s}(nV_1 - V_2) \tag{3.6}$$

The averaged charging power is

$$P = \frac{nV_1 D^2}{4L_s f_s}(nV_1 - V_2) \tag{3.7}$$

Solving for D in Equation 3.7 and substituting into Equation 3.6,

$$I_{\max} = \frac{nP\sqrt{nV_1 - V_2}}{\sqrt{nV_1 L_s f_s}} \tag{3.8}$$

When the phase shift increases, the operation will reach continuous mode as shown in Figure 3.14b. The boundary of the continuous mode and discontinuous mode is given by

$$D = \frac{V_2}{nV_1} < 1 \tag{3.9}$$

When the system is operating in continuous mode,

$$I_{max} = \frac{(nV_1 - V_2)}{4L_s f_s} \frac{(nV_1 D + V_2)}{V_1} \tag{3.10}$$

$$P = \frac{2D(nV_1)^2 - V_2^2 - (nV_1 D)^2}{8 f_s L_s} \frac{V_2}{nV_1} \tag{3.11}$$

Especially when $D = 1$, the maximum power delivered by the system is

$$P = \frac{(nV_1 - V_2)(nV_1 + V_2)}{8 f_s L_s} \frac{V_2}{nV_1} \tag{3.12}$$

and the maximum current on the MOSFETs is

$$I_{max} = \frac{nV_1 - V_2}{4 f_s L_s} \frac{nV_1 + V_2}{V_1} \tag{3.13}$$

Therefore this shows that the peak current of the primary side is a comprehensive function of many parameters of the circuit, such as the leakage inductance of the transformer, the battery voltage, and the switching frequency of the MOSFETs. For a MOSFET-based charger, the stray inductance in the commutating loop could be minimized by realizing a compact design. The primary DC-bus voltage is 150 V if the input voltage is 110 V AC. Therefore, the possibility of overvoltage across the MOSFETs could be negligible. Current impact happens to be a primary concern. Assuming the MOSFETs can handle infinitely large current, the maximum output power of the system is plotted in Figure 3.15 (curves C_1 and C_3) when the battery voltage ranges from 20 to 365 V. The two curves (C_1 and C_3) are for $V_1 = 140$ and 150 V, respectively, which considers the possible fluctuation of the DC-bus voltage.

However, with the decrease in V_2, the risk of a large current impact will be imposed on the MOSFETs if the system is operated under the high-power output. Decreasing the charging power will help reduce the current impact but increases the charging time. Therefore, if the fast charger is regarded as a black box, the output power should be maximized as long as the MOSFET voltage and current are within its capability. In this device, the maximum output power of the system is shown in Figure 3.15 as curves C_2 and C_4 with $V_{dc} = 140$ and 150 V, respectively.

The maximum power that this device can generate is shown as the shaded region in Figure 3.15a. When the battery voltage equals 310 V, the maximum output power can reach 4.5 kW. After dividing the power by the battery voltage, the recommended charging current curve is as detailed in Figure 3.15b,

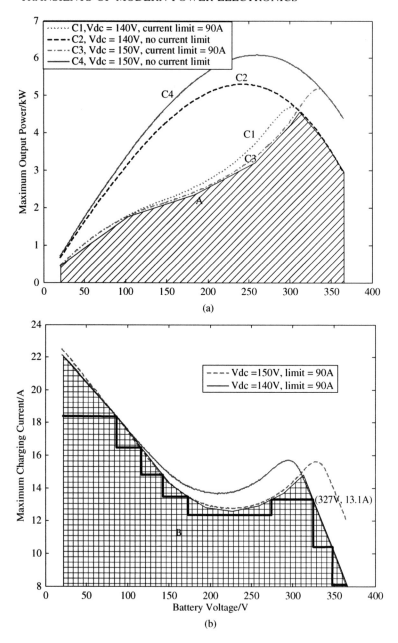

Figure 3.15   Analysis of power capability at different load conditions: (a) power capability of the system and (b) quasi-dynamic system-level SOA.

where the horizontal axis is the battery voltage and the vertical axis is the recommended battery charging current. Any point located inside the grid region will ensure the system works reliably. For simplicity, the control could also be adjusted as shown by the bold lines, that is, in each voltage segment the battery is charged with a different current value.

3.7 Soft-switching control and its application in high-power converters

For switched-mode power supplies, there has been an ever-increasing demand for raising the switching frequency to allow the use of smaller passive components, such as inductors and capacitors, thus providing higher power density and better dynamic performance. The higher switching frequency results in increased switching losses. During the turn-on process, the switching loss is caused mainly by an abrupt change in the energy stored in the parasitic capacitances of the solid state devices. During the turn-off process, the switching loss is mainly caused by the stray inductance. When an active switch is turned off, a voltage spike is induced by the sharp di/dt across the stray inductance. Furthermore, the conventional hard-switching converters exhibit high dv/dt and di/dt in their operation, which causes excessive EMI. This EMI can easily affect the operation of other components and equipment if it is not appropriately controlled.

In the 1980s, soft-switching techniques emerged in converter design practice. Soft switching utilizes resonance theory and control algorithms to trigger the semiconductor switches when the voltage or current equals zero, mostly through resonance of the circuit. The stray inductance and parasitic capacitance cooperate with each other in the resonant process, which suppress the voltage spike in the turn-off stage and in-rush current at the moment of turn-on. If the switch is triggered on or off when the current or voltage is zero, this is defined as zero current switching or zero voltage switching (ZCS/ZVS) [21, 22].

3.7.1 Case 4: ZCS in dual-phase-shift control

In Figure 3.11, if the duty ratio of gate signals is fixed at 50% and the time sequence is as shown in Figure 3.16, then the current waveform can also be changed. It is then possible to realize soft switching. In [23], this control is defined as dual-phase-shift control. The switching process are described below:

$t = t_0$: Q_1 turns on and Q_2 turns off. Q_2 is hard switched off while Q_1 is ZCS on since current flows through the anti-parallel diode, not Q_1.

$t = t_1$: Q_8 turns on and Q_7 turns off. Q_7 is hard switched off while Q_8 is ZCS on.

$t = t_2$: Q_5 turns on and Q_6 turns off. Q_6 is hard switched off while Q_5 is ZCS on.

$t = t_3$: Q_3 turns on and Q_4 turns off. Q_4 is hard switched off while Q_3 is ZCS on.

$t = t_4$: Q_2 turns on and Q_1 turns off. Q_1 is hard switched off while Q_2 is ZCS on.

$t = t_5$: Q_7 turns on and Q_8 turns off. Q_8 is hard switched off while Q_7 is ZCS on.

$t = t_6$: Q_6 turns on and Q_5 turns off. Q_5 is hard switched off while Q_6 is ZCS on.

$t = t_7$: Q_4 turns on and Q_3 turns off. Q_3 is hard switched off while Q_4 is ZCS on.

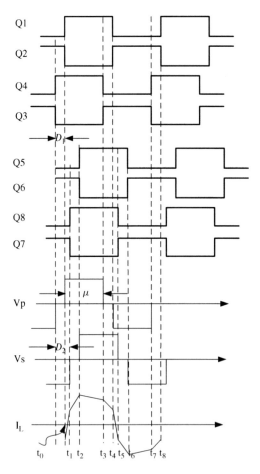

Figure 3.16   Time sequence of dual phase shift. © *[2008] IEEE. Reprinted, with permission, from IEEE Transactions on Industrial Electronics.*

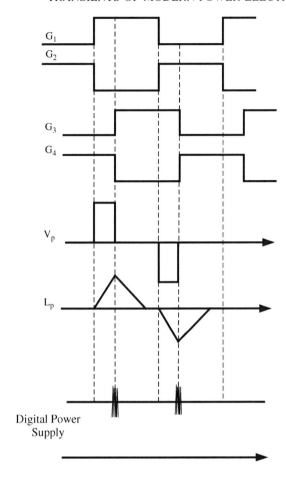

Figure 3.17   Hard-switching control.

Therefore all the turn-off actions are hard switched while all the turn-on actions are ZCS. In the ZCS-on process, all the loss happens at the anti-parallel diode, not in the main switches. However, this analysis is only for specific operating conditions. When the load changes, the condition of soft switching may disappear, as pointed out in [20].

3.7.2   Case 5: Soft-switching vs. hard-switching control in the EV charger*

Consider the unidirectional battery charger shown in Figure 3.13 as an example. T_1 and T_2 are turned on/off with zero current, therefore the switching actions of these two devices can be regarded as soft switching. On the contrary, T_3 and T_4 are switched at the peak current, therefore being hard switched. Hard-switching actions generate large EMI which affects other devices, such as the digital power supply (5 or 3.3 V) as shown in Figure 3.17. With the increase in voltage and current, the amplitude of the noise on the power supply also increases, which interferes with the operation of the core control chips.

References

1. ABB Semiconductors (2003) Reverse Conducting Integrated Gate-Commutated Thyristor 5SHX 08F4502, ABB Switzerland Ltd.
2. Kraus, R. and Hoffmann, K. (1993) *An Analytical Model of IGBTs with Low Emitter Efficiency*, Siemens, Munich, pp. 30–34.
3. Yuan, L., Zhao, Z., and Li, C. (2003) The optimization of snubbers for IGCT-based voltage source inverters. The 29th Annual Conference of the IEEE Industrial Electronics Society, pp. 679–682.
4. Bai, H., Zhao, Z., Eltawil, M., and Yuan, L. (2007) Optimization design of high-voltage-balancing circuit based on the functional model of IGCT. *IEEE Transactions on Industrial Electronics*, **54** (6), 3012–3021.
5. Kraus, R. and Hoffmann, K. (1992) *A Precise Model for the Transient Characteristics of Power Diodes*, Siemens, Munich, pp. 863–869.
6. Rong, Y., Zhengming, Z., and Liqiang, Y. (2008) Fast recovery diode model for circuit simulation of high voltage high power three level converters. *Transactions of China Electrotechnical Society*, **23** (7), 62–67 (in Chinese).
7. Castagno, S., Curry, R.D., and Loree, E. (2006) Analysis and comparison of a fast turn-on series IGBT stack and high-voltage-rated commercial IGBTs. *IEEE Transactions on Plasma Science*, **34** (5), 1692–1696.
8. Qin, Z. and Narayanan, E.M.S. (1997) A novel multi-channel approach to improve LIGBT performance. IEEE International Symposium on Power Semiconductor Devices and ICs, pp. 313–316.
9. Otsuki, M., Kirisawa, M., and Sakurai, K. (1996) A study on current handling capability of dual gate MOS thyristor (DGMOS). International Symposium on Power Semiconductor Devices and ICs, pp. 137–140.
10. Cai, J., Lo, K.F., and Sin, J.K.O. (1999) A latch-up immunized lateral trench-gate conductivity modulated power transistor. International Symposium on the Physical and Failure Analysis of Integrated Circuits, pp. 168–172.
11. ABB Semiconductors (2001) An Application-Specific Asymmetric IGCT, http://www05.abb.com/global/scot/scot256.nsf/veritydisplay/2ad9ae46b17d0216c1256b9d004290a5/$File/Ch01ts.pdf (accessed March 24, 2011).
12. Zhao, Z., Bai, H., and Liqiang, Y. (2007) Transient of power pulse and its sequence in power electronics. *Science in China Series E: Technological Sciences*, **50** (3), 351–360.
13. Clarke, R.C., Brandt, C.D., Sriram, S. *et al.* (1998) Recent advances in high temperature, high frequency SiC devices. Devices and Sensors Conference on High-Temperature Electronic Materials, pp. 18–28.

14. Mihaila, A., Udrea, F., Rashid, S.J. *et al.* (2006) High temperature characterization of 41-SiC normally-on vertical JFETs with buried gate and buried field rings. IEEE International Symposium on Power Semiconductor Devices and ICs, pp. 1–4.

15. Wikipedia (2004) http://en.wikipedia.org/wiki/File:Jfet.png (accessed 24 March, 2011).

16. Zhao, J.H., Tone, K., Li, X. *et al.* (2004) 3.6 mΩ cm^2, 1726 V 4H-SiC normally-off trenched-and-implanted vertical JFETs and circuit applications. *IEE Proceedings: Circuits, Devices and Systems*, **151** (3), 231–237.

17. Wang, Y., Cass, C.J., Chow, T.P. *et al.* (2006) SPICE model of SiC JFETs for circuit simulations. IEEE Workshops on Computers in Power Electronics, pp. 212–215.

18. Bai, H., Pan, S., Mi, C. *et al.* (2010) A functional model of silicon carbide JFET and its use in the analysis of switching-transient and impact of gate resistor, miller effect, and parasitic inductance. *International Journal of Power Electronics*, **2** (2), 164–175.

19. Zhao, Z., Bai, H., and Liqiang, Y. (2008) Analysis of cutting-edge techniques in the high voltage and high power adjustable speed drive systems. *Science in China Series E: Technological Sciences*, **52** (2), 422–449.

20. Mi, C., Bai, H., Wang, C., and Gargies, S. (2008) The operation, design, and control of dual H-bridge based isolated bidirectional DC-DC converter. *IET Power Electronics*, **1** (3), 176–187.

21. Chu, C.-L. and Chen, Y. (2009) ZVS-ZCS bidirectional full-bridge DC-DC converter. International Conference on Power Electronics and Drive Systems, pp. 1125–1130.

22. Bhajana, V. and Reddy, S.R. (2009) A novel ZVS-ZCS bidirectional DC-DC converter for fuel cell and battery application. International Conference on Power Electronics and Drive Systems, pp. 12–17.

23. Bai, H. and Mi, C. (2008) Eliminate reactive power and increase system efficiency of isolated bidirectional dual-active-bridge DC–DC converters using novel dual-phase-shift control. *IEEE Transactions on Power Electronics*, **23** (6), 2905–2914.

4

Power electronics in electric and hybrid vehicles

4.1   Introduction of electric and hybrid vehicles

Modern society relies heavily on fossil fuel-based transportation for economic and social development – freely moving goods and people. There are about 800 million cars in the world and about 250 million motor vehicles on the road in the United States, according to the US Department of Transportation's estimate. In 2009, China overtook the United States to become the world's largest auto maker and auto market, with output and sales respectively hitting 13.79 and 13.64 million units in 2010. With further urbanization, industrialization, and globalization, the trend of rapid increase in number of personal automobiles worldwide is inevitable. The issues related to this trend become evident because transportation relies greatly on oil. Not only are the oil resources on Earth limited, but also the emissions from burning oil products have led to climate change, poor urban air quality, and political conflicts.

Therefore, future personal transportation should provide enhanced freedom, sustainable mobility, and sustainable economic growth and prosperity for society. In order to achieve these, vehicles driven by electricity from clean, secure, and smart energy are essential. Vehicles driven electrically provide many advantages and challenges. Electricity is more efficient than combustion processes in a car. Well-to-wheel studies show that, even if the electricity is generated from petroleum, the equivalent miles that can be driven by 1 gallon (US; 3.79 l) of gasoline is 108 miles (equivalent; 173 km) in an electric car, compared to 33 miles (53 km) in a car driven by an internal combustion engine. Electricity can be generated through renewable sources, such as hydroelectric, wind, solar, and biomass. On the other hand, the current electric grid has capacity available

Transients of Modern Power Electronics, First Edition. Hua Bai and Chris Mi.
© 2011 John Wiley & Sons, Ltd. Published 2011 by John Wiley & Sons, Ltd.

at nights when the use of electricity is off peak. It is ideal to charge electric vehicles (EVs) at night when the grid has extra energy capacity available.

High cost, limited driving range, and long charging time are the main challenges for battery-powered electric vehicles. Hybrid electric vehicles (HEVs), which use both an internal combustion engine and an electric motor to drive the vehicle, overcome the cost and range issues of a pure EV without the need for plugging in to charge. The fuel consumption of HEVs can be significantly reduced compared to conventional gasoline engine-powered vehicles. However, the vehicle still operates on gasoline/diesel fuel.

Plug-in hybrid electric vehicles (PHEVs) are equipped with a larger battery pack and a larger-sized motor compared to HEVs. PHEVs can be charged from the grid and driven over limited distances (20–40 miles) using electricity, referred to as charge-depletion (CD) mode operation. Once the battery energy has been depleted, it operates similar to a regular HEV, referred to as charge-sustain (CS) mode operation or extended-range operation. Since most personal vehicles are for commuting and 75% of them are driven only 40 miles or less daily, a significant amount of fossil fuel can be displaced by deploying PHEVs capable of 40 miles of purely electric propulsion-based driving range. In extended-range operation, a PHEV works similar to a HEV by using the onboard electric motor and battery to optimize the engine and vehicle system operation to achieve higher fuel efficiency. Due to the larger battery power and energy capacity, PHEVs can recover more kinetic energy during braking, thereby further increasing fuel efficiency.

In modern EVs, HEVs, and PHEVs, one of the key technologies for increasing the fuel efficiency is regenerative braking [1]. During regenerative braking, the battery pack is charged by the vehicle's kinetic energy during the braking process instead of converting it into heat as is the case in conventional brakes. Besides batteries, ultracapacitors (UCs) [2] and fuel cells [3] are also common choices. In some cases, batteries, UCs, and fuel cells are even integrated inside the same vehicle [4, 5].

4.2   Architecture and control of HEVs

There are a number of different types of HEVs: series [6], parallel [7], and fuel cell [8]. In a series hybrid vehicle, the internal combustion engine (ICE) drives a generator to produce electricity which in turn is used to drive the electric motor to propel the vehicle. A battery pack is used to store the excessive energy of the engine/generator as well as the energy from regenerative braking. The electric motor drives the wheels through a gearbox. Therefore, from the point of view of the wheel, only the electric motor is used to drive the vehicle, and the transmission of power from both the ICE and battery to the drive wheels is electrical, as shown in Figure 4.1a.

In the parallel HEV shown in Figure 4.1b, the two power sources (ICE and electric motor) are mechanically coupled to drive the wheels. There is also another

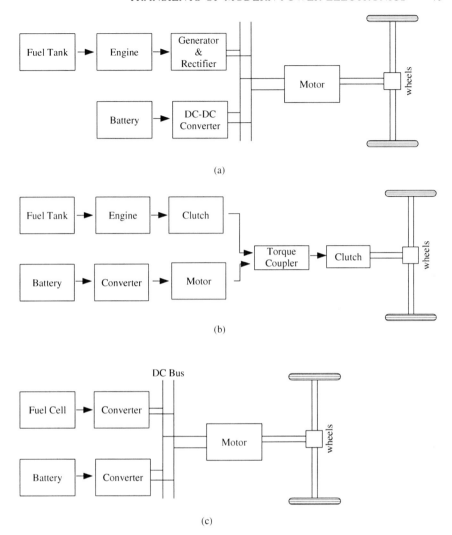

Figure 4.1 Main types of the hybrid electric vehicles: (a) series hybrid electric vehicle, (b) parallel hybrid electric vehicle, and (c) fuel cell electric vehicle.

type of vehicle called the fuel cell vehicle, which uses a fuel cell as the main power source, as shown in Figure 4.1c.

4.3 Power electronics in HEVs

Power electronics is one of the enabling technologies propelling the shift from conventional gasoline/diesel engine-powered vehicles to electric, hybrid, and fuel cell vehicles. Many types of power electronics converters are used in EVs, HEVs,

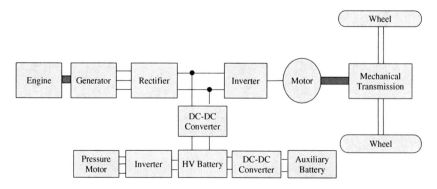

Figure 4.2   Various types of power electronics converters used in a hybrid vehicle system.

and PHEVs, as shown in Figure 4.2 for a series hybrid vehicle. Rectifiers, DC–DC converters, and inverters are all used in the vehicle propulsion system. Auxiliary units, such as pressure pumps, air-conditioning unit, and auxiliary battery, will need inverters or DC–DC converters at a lower power rating. For PHEVs, there is also a battery charger installed on the vehicle or in the charging station.

4.3.1   Rectifiers used in HEVs

Rectifiers are used to convert an AC input to a DC output. Even though controlled rectifiers exist, they are rarely used in automotive applications. Therefore, we will only discuss uncontrolled passive rectifiers and their unique aspects in HEV applications.

4.3.1.1   Ideal rectifier

Figure 4.3 shows a single-phase rectifier and a three-phase ideal rectifier operating from ideal voltage sources. With ideal diode characteristics, the output of a single-phase rectifier can be expressed as follows:

$$V_o = \frac{1}{T/2} \int_0^{T/2} \sqrt{2} V_i \sin(\omega t)\, dt = 0.9 V_i \tag{4.1}$$

where V_o is the output voltage, V_i is the root mean square (RMS) value of the input voltage, T is the period of the input voltage, and ω is the angular frequency of the input.

The output of an ideal three-phase rectifier is

$$V_o = \frac{1}{\pi/3} \int_{-\pi/6}^{\pi/6} (V_a - V_b)\, dt = \frac{1}{\pi/3} \int_{-\pi/6}^{\pi/6} \sqrt{2} V_{LL} \cos(\omega t)\, dt = 1.35 V_{LL} \tag{4.2}$$

where V_{LL} is the line-to-line voltage.

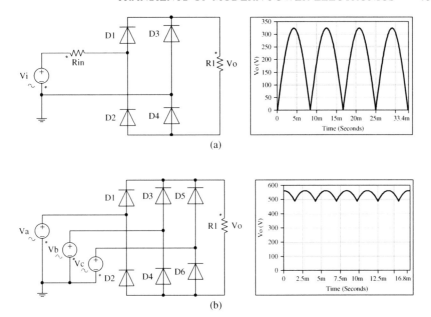

Figure 4.3   Ideal rectifiers: (a) single-phase rectifier circuit and output voltage waveforms; (b) three-phase rectifier circuit and output voltage waveforms.

4.3.1.2   Practical rectifier

In HEV applications, the input to a rectifier is usually the output of a synchronous generator (such as in a series HEV or complex HEV), or an alternator (in a belt–alternator–starter HEV). The circuit and output voltage waveforms of a practical HEV rectifier are shown in Figure 4.4. The generator impedance is in series with the voltage source, and the voltage drop of the diodes is also included. It can be seen that there is a significant voltage drop in a practical rectifier when compared to an ideal rectifier. The voltage drop is caused by the impedance of the generator, which is generally not negligible, different from that of rectifiers connected to an infinite AC grid. In addition, there will be commutation loss due to the inductance of the generator. Therefore, the output voltage can be significantly different between no-load and loaded conditions. The difference is defined as voltage regulation (Figure 4.5).

4.3.1.3   Single-phase rectifier

We will use a single-phase circuit to analyze the voltage regulation, voltage ripple, and commutation.

At no-load conditions, due to the existence of the output capacitor, the output voltage will equal the peak of the input voltage, that is, $V_o = \sqrt{2}V_a = 1.414V_a$. When the load current increases, the impedance of the generator and the diodes

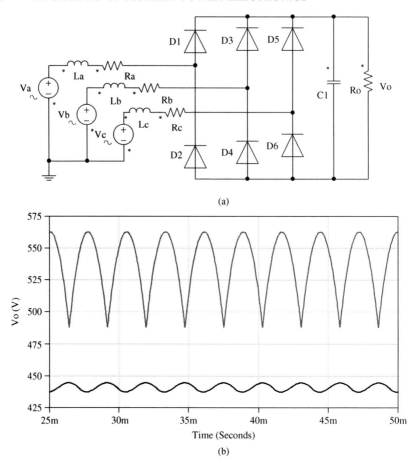

(a)

(b)

Figure 4.4   Practical rectifiers used in HEV. (a) Rectifier circuit. (b) Output voltage in comparison with ideal rectifier: upper curve, ideal rectifier; lower curve, practical rectifier.

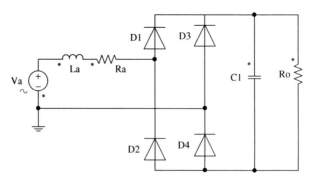

Figure 4.5   Voltage regulation, commutation of practical rectifier used in HEV.

will have a voltage drop on them. If the DC-link capacitor is sufficiently large, then we can assume that the output voltage V_1 (DC-link voltage) is constant. We further assume that diode voltage drop V_D is also a constant.

By solving $\sqrt{2}V_a \sin \omega t = 2V_D + V_1$ we get,

$$\theta_1 = \omega t_0 = \sin^{-1} \left(\frac{2V_D + V_1}{\sqrt{2}V_a} \right) \tag{4.3}$$

The analysis can be divided into two scenarios: discontinuous mode and continuous mode. In discontinuous mode, the AC side current is not continuous. The AC side current starts from zero to build up when $\omega t \geq \theta_1$; the current reaches a maximum when $\sqrt{2}V_i \sin \omega t = 2V_D + V_o$ ($\omega t = \pi - \theta_1$); the current then drops to zero at θ_2, $\theta_2 < \pi + \theta_1$.

Continuous mode   In continuous mode, the AC side current does not reach 0 at $\pi + \theta_1$. In other words, $\theta_2 > \pi + \theta_1$.

In continuous mode, the voltage equation is

$$V_i - L_a \frac{di}{dt} - R_a i - 2V_D = V_1, \quad i(t_0) = 0$$

$$\frac{di}{dt} + \frac{R_a}{L_a} i = \frac{1}{L_a}(\sqrt{2}V_a \sin \omega t - 2V_D - V_1) \quad \text{when} \quad \omega t \geq \theta_1 \tag{4.4}$$

Note that the AC input will not have current until the voltage exceeds the output voltage plus the diode drop. However, the current will continue to flow until it reaches zero at angle θ_2.

We further neglect the resistance. The above differential equation can be simplified and the following solution obtained:

$$i(t) = -\frac{\sqrt{2}V_a}{\omega L_a} \cos \omega t - \frac{2V_D + V_1}{\omega L_a} \omega t + C, \quad \theta_1 \leq \omega t \leq \theta_2 \tag{4.5}$$

Since $i(\theta_1) = 0$, from this equation we can get the expression for C,

$$C = \frac{\sqrt{2}V_a}{\omega L_a} \cos \theta_1 + \frac{2V_D + V_1}{\omega L_a} \theta_1 \tag{4.6}$$

Therefore,

$$i(t) = -\frac{\sqrt{2}V_a}{\omega L_a}(\cos \omega t - \cos \theta_1) - \frac{2V_D + V_1}{\omega L_a}(\omega t - \theta_1), \quad \theta_1 \leq \omega t \leq \theta_2 \tag{4.7}$$

To find θ_2,

$$\frac{1}{\omega L_a} \int_{\theta_1}^{\theta_2} (\sqrt{2}V_a \sin \omega t - 2V_D - V_1) d\omega t = 0 \tag{4.8}$$

$$\cos\theta_2 + \frac{2V_D + V_1}{\sqrt{2}V_a}\theta_2 = \cos\theta_1 + \frac{2V_D + V_1}{\sqrt{2}V_a}\theta_1 \tag{4.9}$$

The output power of the rectifier must be equal to the power consumed by the load:

$$P = \frac{1}{\pi}\int_{\theta_1}^{\theta_2}\sqrt{2}V_a\sin\omega t^* i(t)d(\omega t) = \frac{V_1^2}{R} \tag{4.10}$$

so the output voltage of the rectifier is

$$V_o = \sqrt{\frac{R}{\pi}\int_{\theta_1}^{\theta_2}\sqrt{2}V_a\sin\omega t^* i(t)d(\omega t)} \tag{4.11}$$

This expression cannot be directly solved due to the fact that θ_1 and θ_2 are functions of V_o. But it can be seen that the rectifier output is closely related to the impedance of the generator.

The voltage regulation can then be calculated:

$$\Delta V_o = \frac{V_1 - V_o}{V_o} \times 100\% \tag{4.12}$$

Apparently, the voltage regulation is a function of the internal impedance of the generator and the output power.

4.3.1.4 Voltage ripple

The above derivation assumes that the output voltage is constant. However, it can be seen that the current from the AC input is discontinuous. This means that during the portion of the cycle when there is no current from the AC side (e.g., from 0 to θ_1, and from θ_2 to π if $\theta_2 < \pi$), the load current is supplied by the capacitor.

However, due to the nonlinearity of the current, when the load current is less than the AC side current, the capacitor still has to supply some current to the load. Therefore, we can assume that if the load current is constant, and that the capacitor supplies current to the load 50% of the time, then the voltage ripple is

$$\Delta V_o = \frac{1}{2}\frac{1}{C}\frac{V_o}{R}\frac{\pi}{\omega} \tag{4.13}$$

We take the average of the above two equations to get the average voltage ripple.

Discontinuous mode   In discontinuous mode, the AC side current starts at $\omega t = \theta_2 > \theta_1$, and drops to zero at $\omega t > \pi + \theta_1$:

$$\frac{di}{dt} + \frac{R_a}{L_a}i = \frac{1}{L_a}(\sqrt{2}V_a\sin\omega t - 2V_D - V_1), \quad \theta_2 \leq \omega t \leq \theta_2 + \pi \tag{4.14}$$

and

$$i(\pi - \theta_2) = i(\theta_2) = 0 \tag{4.15}$$

When neglecting R_a,

$$i(t) = -\frac{\sqrt{2}V_a}{\omega L_a}(\cos\omega t - \cos\theta_2) - \frac{2V_D + V_1}{\omega L_a}(\omega t - \theta_2), \quad \theta_2 \le \omega t \le \pi + \theta_2 \tag{4.16}$$

To find θ_2, let

$$\frac{1}{\omega L_a}\int_{\theta_2}^{\pi+\theta_2}(\sqrt{2}V_a\sin\omega t - 2V_D - V_1)d\omega t = 0$$

$$\theta_2 = \cos^{-1}\left[\frac{\pi(2V_D + V_1)}{2\sqrt{2}V_a}\right] \tag{4.17}$$

The boundary condition is $\theta_1 = \theta_2$. Therefore, the boundary condition happens when

$$2V_D + V_1 = \frac{2\sqrt{2}V_a}{\sqrt{\pi^2 + 4}} \tag{4.18}$$

The above analysis is based on a single-phase generator. Since most generators and motors are three phase, it is worth looking at three-phase circuits.

Again, if the output capacitor is sufficiently large, then we can assume that the output voltage is constant. The diodes only conduct $60°$ in each cycle. The voltage equation can be written as

$$V_{LL} - 2L_a\frac{di}{dt} - 2R_a i - 2V_D = V_1, \quad -\frac{\pi}{6} \le \omega t \le \frac{\pi}{6} \tag{4.19}$$

where V_{LL} is the line-to-line voltage and V_1 is the DC-link voltage with load. This equation can be solved using the same method for the single-phase analysis.

4.3.2 Buck converter used in HEVs

4.3.2.1 Operating principle

A buck converter will step down a higher voltage DC input to a lower voltage DC output. The typical application of a buck converter in a HEV is to step down the hybrid battery voltage (typically 200–400 V) to charge the auxiliary battery (14 V). The uniqueness is the large difference between the input and output voltage of the converter and the small duty ratio (3.5%) needed to control the switching. Figure 4.6 shows the main circuit of a buck converter. It consists of a switch, a freewheeling diode, and an LC filter.

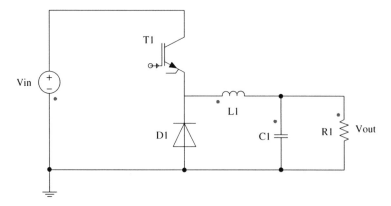

Figure 4.6   A buck converter.

The small duty ratio will make control and regulation very difficult. It also affects the design of the inductor, capacitor, current ripple, and voltage ripple. As a starting point of the analysis, we assume that the components are ideal, that is, the voltage drop is zero when turned on. We further assume that the output voltage is constant. When the switch is turned on, the voltage across the inductor is

$$V_L = V_d - V_o \tag{4.20}$$

When the switch is turned off, we assume the circuit is operating in continuous mode, so the voltage drop across the inductor is

$$V_L = -V_o \tag{4.21}$$

In steady state operation, the average voltage of the inductor must be zero. Therefore,

$$(V_d - V_o)DT_s = V_o(1 - D)T_s \tag{4.22}$$

and

$$V_o = DV_d \tag{4.23}$$

The current ripple in the inductor is

$$\Delta I_L = \frac{1}{L}V_o(1 - D)T_s \tag{4.24}$$

The voltage ripple of the output can be calculated. We assume that the load current is constant, so all the current ripple will enter the capacitor:

$$\Delta V_o = \frac{1}{C}\frac{1}{2}\frac{\Delta I_L}{2}\frac{T_s}{2} = \frac{T_s^2}{8LC}V_o(1 - D) \tag{4.25}$$

4.3.2.2 Nonlinear model

The above analysis is based on assumptions that the output voltage is relatively constant and the parasitic parameters (resistance, inductance) can be neglected. Due to the fact that the ratio of the input and output voltages is very large, these assumptions may not hold true.

In order to accurately analyze the relationship and the influence of various parameters, we can use a detailed model to describe the system. In continuous mode, when the switch is turned on,

$$V_d = r_d i_L + r_s i_L + L\frac{di_L}{dt} + r_L i_L + V_o \qquad (4.26)$$

$$i_L = i_c + i_o = C\frac{dV_o}{dt} + i_o = C\frac{dV_o}{dt} + \frac{V_o}{R}$$

where r_d, r_s, and r_L are the equivalent resistance of the diode, switch, and inductor, respectively; and i_o, i_c, and i_L are the current through the load resistance, the capacitor, and the inductor, respectively. The above equation can be rewritten as

$$\begin{bmatrix} \dfrac{di_L}{dt} \\ \dfrac{dV_o}{dt} \end{bmatrix} = \begin{bmatrix} -\dfrac{r_d + r_s + r_L}{L} & -\dfrac{1}{L} \\ \dfrac{1}{C} & -\dfrac{1}{CR} \end{bmatrix} \begin{bmatrix} i_L \\ V_o \end{bmatrix} + \begin{bmatrix} \dfrac{1}{L} \\ 0 \end{bmatrix} V_d \qquad (4.27)$$

When the switch is closed,

$$\begin{bmatrix} \dfrac{di_L}{dt} \\ \dfrac{dV_o}{dt} \end{bmatrix} = \begin{bmatrix} -\dfrac{r_D + r_L}{L} & -\dfrac{1}{L} \\ \dfrac{1}{C} & -\dfrac{1}{CR} \end{bmatrix} \begin{bmatrix} i_L \\ V_o \end{bmatrix} \qquad (4.28)$$

These equations can be solved using numeric tools such as MATLAB.

4.3.3 Non-isolated bidirectional DC–DC converter

The bidirectional DC–DC converter in a HEV is also sometimes called an energy management converter or boost DC–DC converter. This DC–DC converter is a high-power converter that links the high-voltage (HV) battery at a lower voltage with the HV DC bus. The typical voltage of a battery pack is designed to be 300–400 V. The best operating voltage for a motor and inverter is around 600 V. Therefore, this converter can be used to match the voltages of the battery system and the motor system. Other functions of this DC–DC converter include optimizing the operation of the powertrain system, reducing the ripple current in the battery, and maintaining DC-link voltage, hence high-power operation of the powertrain.

4.3.3.1 Operating principle

The DC–DC converter provides bidirectional power transfer. The operating principle is shown in Figure 4.7.

Buck operation   In buck operation as shown in Figure 4.7b, the power is transferred from V_d to V_B. When T_1 is closed and T_2 is open, since $V_d > V_B$, $V_L = V_d - V_B$ and the inductor current I_L builds up. When T_1 is open, the inductor current I_L continues to flow through D_2. $V_L = -V_B$.

Assuming ideal components and a constant V_o, the inductor current over one cycle in steady state operation will remain the same, that is,

$$\int_0^{t_{1on}} (V_d - V_o)dt = \int_{t_{1on}}^{t_{1on}+t_{1off}} (-V_o)dt \tag{4.29}$$

$$V_o = \frac{t_{1on}}{T} V_d = D_1 V_d \tag{4.30}$$

where D_1 is the duty ratio defined as the percentage of on-time of switch T_1:

$$D_1 = \frac{t_{1on}}{T} \tag{4.31}$$

Boost operation   In boost operation, the power is transferred from V_B to V_d. When T_2 is closed and T_1 is open, V_B and the inductor form a short circuit through switch T_2 as shown in Figure 4.7c, therefore $V_L = V_B$ and the inductor current I_L builds up. When T_1 is open, the inductor current continues to flow through D_1 to V_d, therefore $V_L = V_d - V_B$:

$$\int_0^{t_{2on}} V_o dt = \int_{t_{2on}}^{t_{2on}+t_{2off}} (V_d - V_o)dt \tag{4.32}$$

$$V_d = \frac{1}{1 - D_2} V_o \tag{4.33}$$

where D_2 is the duty ratio defined as the percentage of on-time of switch T_2:

$$D_2 = \frac{t_{2on}}{T} \tag{4.34}$$

In the control of a bidirectional boost converter, since T_1 and T_2 cannot be switched on simultaneously, a practical control strategy is to turn T_2 off while T_1 is on and vice versa. In this case,

$$D_2 = 1 - D_1 \tag{4.35}$$

4.3.3.2 Maintaining constant torque range and power capability

The above analysis neglected the internal impedance of the battery. In fact, the impedance is often not negligible. When an electric motor and inverter are directly

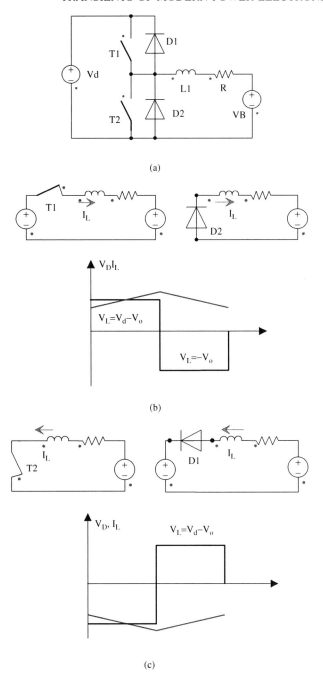

Figure 4.7   Operation of the bidirectional boost converter: (a) circuit topology; (b) inductor voltage and current waveform during buck operation; and (c) inductor voltage and current waveform during boost operation.

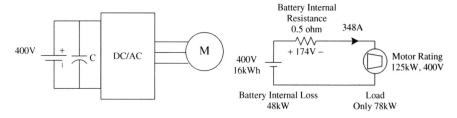

Figure 4.8   Powertrain motor directly connected to battery without the DC–DC converter.

connected to the battery without a bidirectional DC–DC converter as shown in Figure 4.8, as the current (power or torque) goes up, the battery terminal voltage starts to drop because of the voltage drop on the battery internal impedance. For example, a 16 kWh lithium-ion battery with iron phosphate chemistry will have an internal impedance of approximately 0.5 Ω. If the powertrain inverter/motor is rated at 125 kW, 400 V, 90% efficiency, the rated current is 348 A at 400 V. At this current (348A), the battery internal voltage drop is 174 V. This voltage drop will significantly affect the performance of the powertrain motors. In fact, in this example, the maximum power that can be delivered to the motor is only 78 kW. In addition, due to the available voltage at the input, the motor constant torque region is also affected. In the above example, the constant torque region shrinks by 43.5%.

Another factor is that battery voltage is related to the battery state of charge (SOC). As the battery SOC drops, the battery voltage will also drop. Therefore, the available voltage in a motor/inverter terminal is also changed, which will make it difficult to maintain the constant torque range.

When a DC–DC converter is inserted between the battery and inverter/motor as shown in Figure 4.9, the DC-bus voltage before the inverter can be maintained as constant. Therefore, the constant torque range will not be affected by battery SOC or large power draw by the inverter/motor.

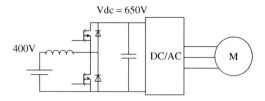

Figure 4.9   Powertrain motor connected to battery through a DC–DC converter.

The above analysis assumes that the battery system is designed to handle large power dissipation during large power draw.

4.3.3.3   Reducing current ripple in the battery

Due to the switching actions of the inverter used in the powertrain system, there are abundant high-frequency current harmonics on the DC side. The amount of current ripple that goes in/out of the battery depends on the switching methodology, switching frequency, and the capacitance on the DC bus. When there is no DC–DC converter in place, the amount of ripple current of the battery is determined by the DC-bus capacitance C and the ratio of capacitor impedance to battery impedance. Without the capacitance, the battery current will be directly determined by the switching status of the DC/AC inverter, that is, the combination of the three-phase current of the motor, as shown in Figure 4.10. When there is a DC-bus capacitor in parallel with the battery, the amount of current ripple flowing in/out of the battery is determined by the capacitance and parasitic impedance of the DC-bus capacitor. For example, if $C = 10\,mF$, the capacitive

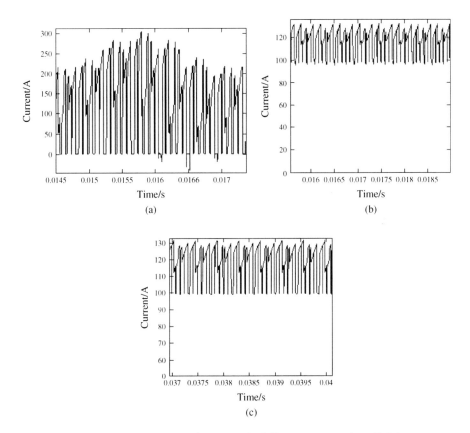

Figure 4.10   *Battery current without a DC–DC converter: (a) no DC-bus capacitance; (b)* $C = 1\,mF$, $R_c = 100\,m\Omega$; *and (c)* $C = 10\,mF$, $R_c = 100\,m\Omega$.

impedance of the capacitor at switching frequency is only 2.65 mΩ, which is far less than the internal impedance of the battery. Ideally the high-frequency ripple will flow through the capacitor and the battery current is supposed to be constant.

However, the equivalent series resistance (ESR) of the capacitor is also not negligible. A high-quality 10 mF capacitor has 26 mΩ internal resistance and the second-class capacitor has 100 mΩ. The quality of the capacitor affects the current ripple of the battery. The lower the capacitor impedance, the lower the battery current ripple, as shown in Figure 4.10. High-frequency current ripple is believed to be harmful to battery life.

When a DC–DC converter is added, the battery current can be maintained with relatively small ripple, as shown in Figure 4.11.

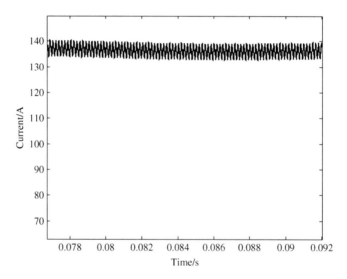

Figure 4.11   Battery current when a DC–DC converter is inserted between the inverter and the battery (I = 137 A with current ripple 5 A).

4.3.3.4 Enhanced regenerative braking

The regenerative braking of the two topologies, that is, one with and one without a DC–DC converter, will also be different. In the topology where there is no DC–DC converter, the DC-bus voltage will fluctuate during transition from motoring to braking. For example, if the motor is initially motoring at 50 kW, and the battery internal voltage is 400 V with 0.5 Ω internal resistance, then the battery current is 155 A and the DC-bus voltage is 322 V. If the motor is switched to braking at 50 kW, then the battery current is 110 A and the DC-bus voltage is 455 V. This dramatic change of DC-bus voltage will make the motor control, such as vector control, very difficult.

On the other hand, in a system that contains a DC–DC converter between the inverter/motor DC bus and the battery, the DC-bus voltage can be maintained relatively constant. Hence, the transition between motoring and braking is easier to handle.

4.3.4   Control of AC induction motors*

To control the speed and/or torque of an induction motor (IM), an inverter is needed. Generally speaking, the topologies for inverters to drive three-phase motors can be classified into two-level and multilevel ones. Compared to the two-level inverter, multilevel ones have many advantages, for example, less voltage–current harmonics and less voltage discontinuity. However, due to the large number of devices, the complexity and size of multilevel inverters also increases compared to two-level inverters. In the domain of motor drives, multilevel inverters, especially three-level inverters, are prevalent. However, in the vehicle domain they are still not widely accepted.

There are various control algorithms for the induction motor [9–11]. These algorithms are similar in terms of steady state characteristics. However, in dynamic processes, there will be some difficulties to overcome.

4.3.4.1   Pre-excitation in the starting process and sneak pulses

In every motor drive system, the starting process is always a very important aspect due to the large in-rush current. Figure 4.12 shows the typical starting current in a 380 V, 160 kW induction motor without proper pre-excitation [12].

The starting current can be up to six times the steady state current. The large starting current is attributed to the fact that the magnetic field of the motor and transformer has not been established completely before start-up. For example, in

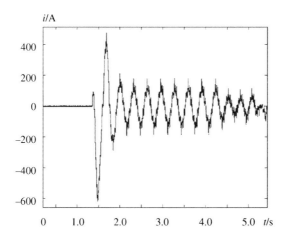

Figure 4.12   Starting in-rush current of one motor drive system.

the case of an induction motor, according to field-oriented control theory [9], the flux linkage and electromagnetic torque can be expressed as follows:

$$\Psi_{rd} = \frac{L_m}{1 + \tau_r p} i_{sd} \tag{4.36}$$

$$T_{em} = p_n \frac{L_m}{L_r} i_{sq} \Psi_{rd} \tag{4.37}$$

Without considering the dynamic process $\tau_r p$, in order to reach the target T_{em}^*, we have

$$i_s = \sqrt{i_{sd}^2 + i_{sq}^2}$$

$$= \sqrt{\left(\Psi_{rd} \frac{1 + \tau_r p}{L_m}\right)^2 + \left(T_{em} \frac{L_r}{p_n L_m \Psi_{rd}}\right)^2} \tag{4.38}$$

$$\geq \sqrt{2 \frac{L_r}{p_n L_m^2} T_{em}^*}$$

The starting current reaches a minimum iff $\Psi_{rd} = \sqrt{L_r T_{em}^* / p_n}$. The back electromotive force would restrain the stator current as long as the magnetic field is fully established. Both excessive and insufficient magnetic flux linkage may cause a large starting current.

In most industrial applications, a good starting performance means not only eliminating the in-rush current, but also generating a large output torque. Hence it is necessary to optimize the relevant parameters of the pre-exciting voltage vector and starting voltage vector.

During normal operation, the flux linkage equations can be expressed as

$$\Psi_S = L_S I_S + L_m I_r \tag{4.39}$$

$$\Psi_r = L_m I_S + L_r I_r$$

Without considering the motor remanent magnetic field and the rotor flux, the stator leakage flux and mutual flux linkage are described as the following when a DC voltage U_S is imposed on the terminals of a squirrel cage induction motor:

$$i_S(t) = \begin{cases} 0 & t < t_0 \\ \dfrac{U_S}{R_S}(1 - e^{-(t-t_0)/\tau}) & t \geq t_0 \end{cases} \tag{4.40}$$

$$\Psi_S(t) = \begin{cases} 0 & t < t_0 \\ \tau U_S(1 - e^{-(t-t_0)/\tau}) & t \geq t_0 \end{cases} \tag{4.41}$$

where τ is the stator time constant and R_S is the stator resistance.

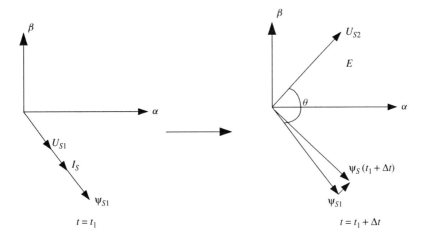

Figure 4.13   The vector scheme when pre-excitation is finished and the motor starts up.

The moment $t = t_1$ is the end of pre-excitation and the initial point of the starting-up process (Figure 4.13). U_{S1} is the exciting vector and U_{S2} is the initial starting vector, as described in the $\alpha\beta$ coordinates in the following vector scheme.

Assuming Δt is small enough, then $I_{S1}(t_S) \approx I_{S2}(t_S)$, hence

$$\frac{d}{dt}\Psi_S = \frac{\Psi_S(t_S + \Delta t) - \Psi_S(t_S)}{\Delta t} = U_{S2} - I_{S1}R_S \qquad (4.42)$$

and Equation 4.42 $\times I_{S2}(t_S)$ is

$$\frac{\Psi_S(t_S + \Delta t) \times I_{S2} - \Psi_S(t_S) \times I_{S2}}{\Delta t} = U_{S2} \times I_{S2} \qquad (4.43)$$

Also

$$T_e = p_n (\Psi_S \times i_S) \qquad (4.44)$$

then

$$\left|\frac{dT_e}{dt}\right| = p_n |U_{S2} \times I_{S2}| \qquad (4.45)$$

Combining with Equations 4.40–4.41, for $t_1 \geq t_0$,

$$\Delta T_e \approx p_n \frac{(1 - e^{-(t_1 - t_0)/\tau})}{R_S} U_{S2} \times U_{S1} \Delta t \qquad (4.46)$$

From Equation 4.46, vectors U_{S1} and U_{S2} must be orthogonal. However, due to the bias magnetic field, remnant flux linkage, and other nonlinear factors, the angle between U_{S1} and U_{S2} might not be exactly $90°$. The optimal combination of the excitation current I_{S1}, the excitation interval $t_1 - t_0$, and the excitation angle θ can be obtained from statistical tests to output a large starting torque with small starting current.

For the two-level inverter, the magnetizing voltage vectors are easier to select. In three-level NPC inverters, vector selection is more complicated, as shown in Figure 4.14a. Here **0NP** and **000**, $\vec{V}_{56}$ and $\vec{V}_{00}$ as shown, are alternatively imposed

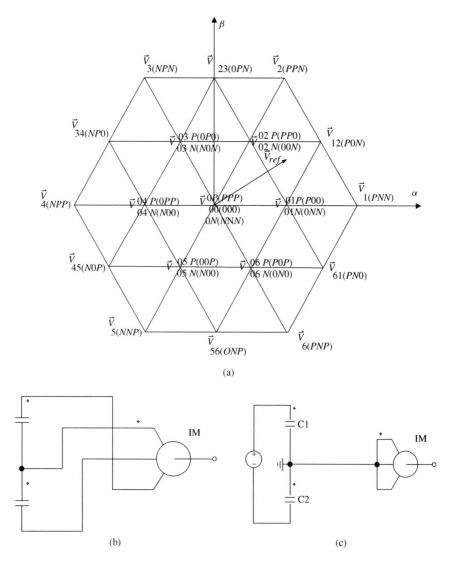

(a)

(b)   (c)

Figure 4.14   (a) Motor model and pre-excitation vector, (b) motor model of **0NP**, and (c) motor model of **000**.

on the transformer terminals to establish the magnetic field of the transformer and the motor before the starting process. Another orthogonal vector, **P00**, $\vec{V}_{01P}$, is used to start the motor.

The middle vector, like **0NP**, and the small vector will influence the neutral point balance [13]. Fortunately, since the motor current is zero before DC pre-excitation, alternating vectors **0NP** and **000** will not impact the balance of the neutral point, as shown in Figure 4.14b,c.

A flowchart is shown in Figure 4.15 for the induction motor pre-excitation. Also, the actual effect of DC pre-excitation is shown in Figure 4.16. Compared to the operation mode without DC pre-excitation whose starting current is over 600 A, the DC pre-excitation reduced the starting current to about 200 A.

Choosing the appropriate voltage vectors to pre-excite the motor is very important for safe operation. However, in this process, some interesting phenomena (possibly dangerous) will likely happen if not controlled properly. We define these as sneak pulses, which are unexpected and small-probability pulses caused by system parasitic parameters in reaction to the control algorithm. These sneak pulses could damage the entire system [14].

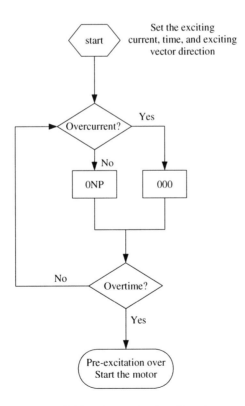

Figure 4.15   Pre-excitation flowchart.

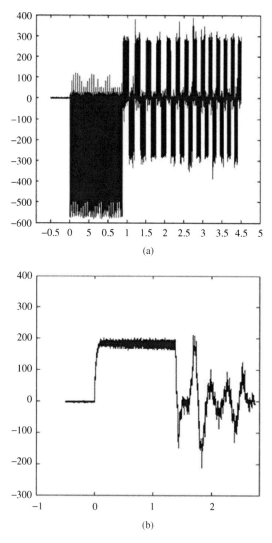

Figure 4.16   The effect of DC pre-excitation: (a) the primary current during DC pre-excitation, and (b) the motor current during DC pre-excitation.

During the DC pre-excitation of the three-level inverter, the vectors **PPP** and **PNP** shown in Figure 4.14a are utilized to establish the DC magnetic field before starting up in order to minimize the current impact on the motor. When the output voltage switches from **N** to **P** for phase **B** as shown in Figure 4.17, although Sb3 and Sb4 are triggered simultaneously, Sb3 might be off prior to Sb4 due to the asynchronous gate signals or the diversity of the semiconductors. This will cause one device to undertake the full DC voltage and destroy the entire bridge, especially in a high voltage inverter.

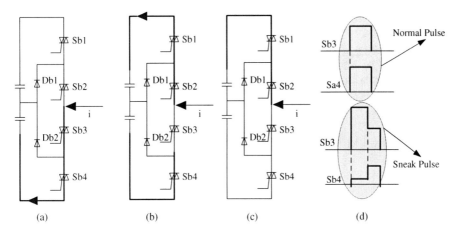

Figure 4.17   The sneak pulse in DC pre-excitation: (a) level N, (b) sneak pulse, (c) level P, (d) Sneak pulse.

The scenario of Figure 4.17b directly results in the voltage imbalance between Sb3 and Sb4, as shown in Figure 4.17d. To eliminate such a sneak pulse, improvements to the algorithm are required. The optimal vectors **0NP** and **000**, instead of **PNP** and **PPP**, have been proven effective in eliminating the above sneak pulse.

Such phenomena also occur during the motor turn-off process. In this process, hypothetically all the semiconductors in the bridges are turned off simultaneously. However, asynchronous gate signals can result in one semiconductor undertaking the full DC-bus voltage as shown in Figure 4.18, where Sa2 is turned off prior to Sa1, Sa3, and Sa4. When Sa2 is turned off with $I_0 > 0$, the current is quickly commutated from Sa1 and Sa2 to the body diodes of Sa3 and Sa4. At that time, Sa1 is still on, which makes Sa2 subject to the full DC-link voltage. This is similar to Figure 4.17.

These sneak pulses are closely related to the control algorithms and circuit topology. For example, if each semiconductor is equipped with a turn-off snubber circuit, such as a RC snubber, this sneak pulse will not be predominant. In two-level DC–DC converters, such a sneak pulse will also not occur.

4.4   Battery chargers for EVs and PHEVs

EVs and PHEVs need to be charged from the electric grid. Hence the battery charger is one of the most critical components in an EV and PHEV. In order to charge the EV and PHEV battery quickly and efficiently, as well as maintain the health of the battery system, an appropriate charger topology and control algorithm are important. There are a number of different charger topologies and algorithms available for this purpose, depending on the requirements of the power

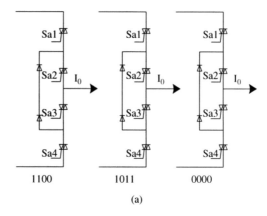

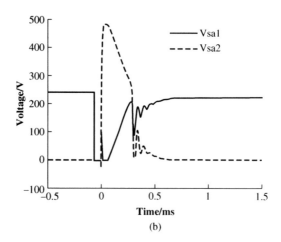

Figure 4.18   Sneak pulse in the pulse-blocking process: (a) sneak pulse in the commutating process, and (b) measured sneak pulse.

rating, whether bidirectional power flow is required, and whether fast charging is necessary. As far as safety is concerned, inductive and wireless chargers may be good alternatives to traditional conductive chargers.

EV and PHEV battery chargers significantly differ from the battery chargers designed for consumer electronics. First, the power level is much higher, ranging from 3 kW to more than 10 kW for a passenger car charger, compared to a few watts or a few tens of watts in consumer electronics. This not only makes the topology essentially different, but also presents concerns with isolation, safety, and EMI issues. Second, a special connection and communication between the charger and the grid are required, referring to the SAE J1772 standard, while consumer electronics can be directly plugged into a 110 or 220 V AC outlet. Third, there are much more stringent requirements for the charger to provide better battery life (more than 10 years required in EVs vs 2–4 years in consumer electronics). Fourth, the voltage level of the battery in an EV and PHEV is much higher (200–400 V), which will require boost conversion, than the low-voltage ratings (a few tens of volts) in consumer electronics, which require buck conversion. Fifth, bidirectional power flow may be required in some EV and PHEV applications, while in consumer electronics the charger is always unidirectional. Lastly, the charger is typically onboard the vehicle and essentially has to experience extreme temperatures, vibrations, and other harsh conditions.

4.4.1   Unidirectional chargers

Most EVs and PHEVs in the near term will be equipped with a unidirectional charger which receives AC power from the electric utility grid and charges the battery onboard the EV or PHEV as shown in Figure 4.19. It consists of a front end rectifier, a power factor correction unit, and an isolated DC-DC converter.

Figure 4.19   Typical composition of an EV and PHEV charger.

As far as topology is concerned, there are a few potential candidates for the main charging circuit, namely, flyback, forward, half bridge, and full bridge.

4.4.1.1   Forward/flyback DC–DC converters

Figure 4.20 shows the circuit and operation of both a forward and a flyback converter, where R is the battery internal resistance, E is the battery internal voltage, and V_o is the output voltage across the battery (including battery internal voltage and voltage drop across the internal resistance).

The operation of the forward converter is similar to the buck converter and the flyback converter is derived from the buck-boost converter [15, 16]. When S_1 in Figure 4.20a or c turns off, the leakage inductance of the transformer will exhaust all the stored energy on the switches, thus inducing significant voltage spikes. Besides a snubber circuit used to mitigate this electrical stress, the third winding of the transformer is needed to demagnetize the remnant magnetic energy when the switch is turned off.

In the flyback converter, when D_1 conducts, the load voltage will be induced to the primary side. Therefore, in the off state of S_1, the voltage across S_1 is $V_{in} + V_o/n$, where n is the turns ratio of the isolation transformer. It indicates that although it may not be necessary to have a filtering inductor in the flyback converter, the semiconductor switch will in fact undergo a higher voltage stress.

4.4.1.2   Half-bridge DC–DC converter

Figure 4.21 shows the half-bridge converter topology where S_1 and S_2 are switched on/off with a phase shift of 180°. The leakage inductance of the

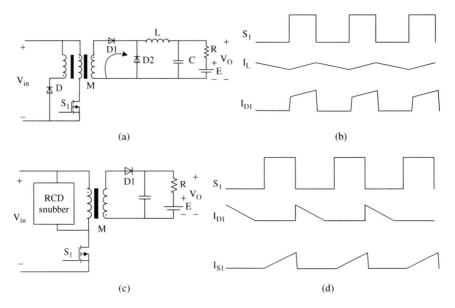

Figure 4.20   Forward/flyback chargers: (a) forward DC–DC converter, (b) oper-ational theory of flyback DC–DC converter, (c) flyback DC–DC converter, and (d) operational theory of flyback DC–DC converter.

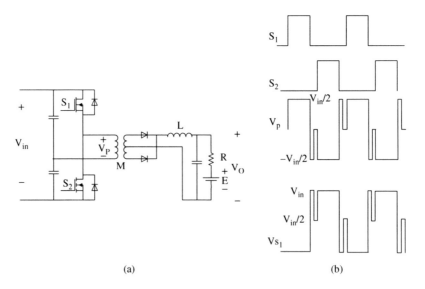

Figure 4.21   Half-bridge topology: (a) half-bridge converter, (b) operational modes.

transformer behaves as the component of energy transfer. Assuming the parasitic inductance of the commutating loop is equal to zero, the voltage spike across the semiconductors will disappear [17].

4.4.1.3   Full-bridge DC–DC converter

Figure 4.22 shows the circuit topology and operation of a full-bridge DC–DC converter. Compared to the half-bridge converter where the primary-side voltage of the transformer is one half of the DC voltage, the full DC voltage is utilized in the full-bridge converter. Similar to the half-bridge converter, the leakage inductance of the transformer in the full-bridge converter does not contribute to any voltage spike across the switches. This leakage inductance should be deployed appropriately [18, 19]. The operation of the full-bridge converter has already been mentioned in chapter 3.

A simulation comparison can be done based on the same input and output parameters, for example, 400 V DC input, 365 V DC output, and 5 kW output

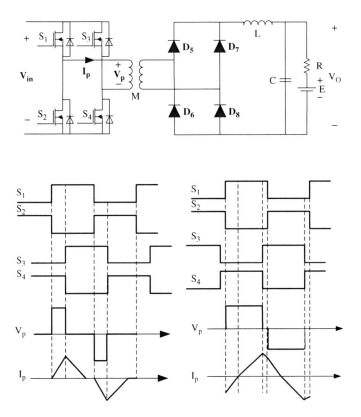

Figure 4.22   Full-bridge topology: (a) full-bridge converter, (b) operation in discontinuous mode, and (c) operation in continuous mode.

power. In the following analysis, the electrical stress and output capacity are compared based on simulation and some experimental results.

4.4.1.4 Voltage stress

The voltage stress of the devices used in each of the above topologies is shown in Figure 4.23, respectively.

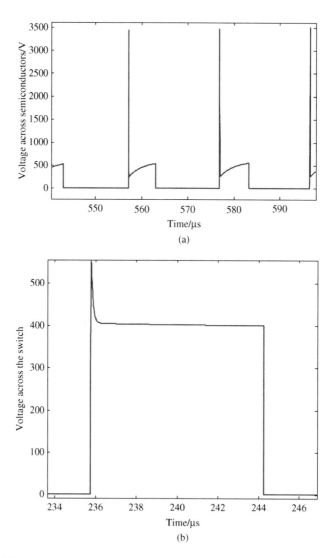

Figure 4.23   Voltage waveforms of one switch equipped in different EV battery chargers: (a) forward converter, no snubber; (b) forward converter, with snubber; (c) flyback converter, no snubber; (d) flyback converter, with snubber; (e) half-bridge, no snubber; (f) full-bridge, no snubber.

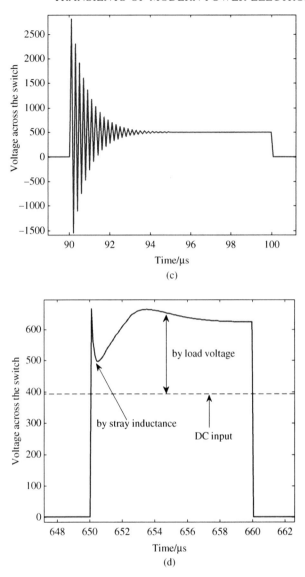

Figure 4.23   (continued)

Without a snubber circuit, the voltage spike is too high for the semiconductors to undertake in the flyback and forward circuits, even with the leakage inductance of the transformer minimized to $2\,\mu H$, as shown in Figure 4.23a,c. It will be different in the case of the half and full bridges, as shown in Figure 4.23e,f. If the DC bus is laid out appropriately, the stray inductance of the commutating loop can be minimized to the nanohenry level, making voltage spikes across the semiconductors negligible. Hence the semiconductors in half- and full-bridge chargers will have less voltage stress than forward and flyback converters.

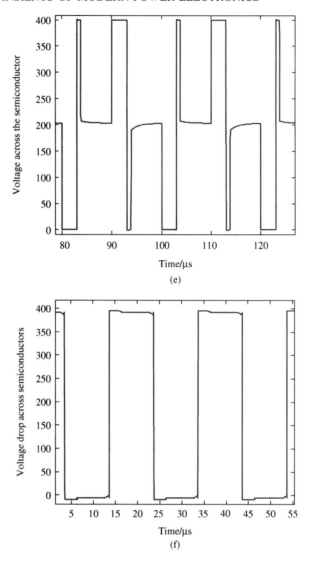

Figure 4.23   (continued)

4.4.1.5   Current stress

For forward chargers, the current ripple on the primary side of the transformer can be limited by the smoothing inductor. For flyback chargers, it can be eliminated by the magnetizing inductance of the transformer. Therefore the current flowing through the switches is nI_o. For half- and full-bridge converters, the primary current is complicated. When the equivalent inductance of the transformer is 30 µH, the simulated current is shown in Figure 4.24, where the filtering inductance of forward converter is always 1 mH.

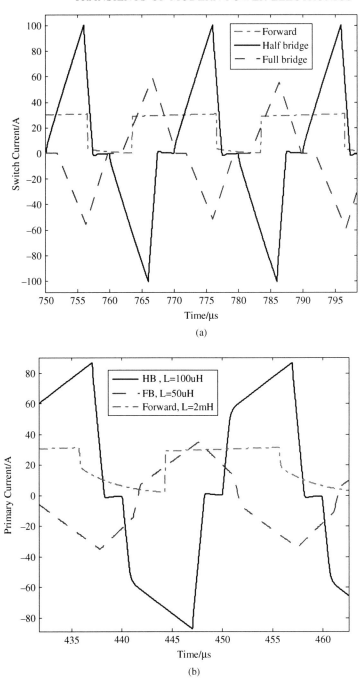

Figure 4.24   Comparison of current stress (a) without filtering inductor and (b) with filtering inductor.

Theoretically, since the primary voltage of the transformer of the half-bridge (HB) converter is only one-half of the DC input, or one-half of that of the forward/flyback or full-bridge (FB) converters, the current of the HB charger is larger than that in the other three topologies at the same output power level. From Figure 4.24a, without any output filtering inductance, the peak current of a FB converter is 60 A and that of a HB charger is 100 A. When a 100 μH filtering inductance is added, the primary current is lowered to 40 A for the FB converter and 90 A for a HB converter.

In this simulation a filtering inductor must exist for a forward converter, otherwise the charging current will not be continuous. For the HB and FB converters, this filtering inductance can be eliminated. However, the primary current of the transformer will increase when no smoothing inductor is used, which means that, with the same semiconductors, their output capability is decreased. When the inductance is increased to 50 μH for the FB converter, the primary current of the transformer is significantly reduced and is comparable to that of the forward converter.

4.4.1.6 Switching losses

Switching losses are another concern in determining which topology is the best. For forward and flyback converters, all the semiconductors are hard switched. Hence, the switching losses are significant. Snubber circuits are the most direct and effective approach to mitigate switching losses. ZVS and ZCS could also be implemented; however, auxiliary semiconductors are needed [20, 21].

For HB and FB converters, soft-switching control is easy to implement. The parasitic capacitance of semiconductor switches and leakage inductance of the transformer could construct a resonant circuit to realize ZVS/ZCS [22].

4.4.1.7 Cost

Compared to forward/flyback and HB converters, FB chargers have the smallest electrical stress for the same input/output parameters. They also have the highest DC-voltage utilization, and therefore smaller current stress than HB converters. In addition, soft switching can be easily realized. However, this topology has the largest number of semiconductor switches. Therefore a cost comparison needs to be addressed. Table 4.1 lists the components needed for the main circuit topologies of different chargers.

A quantitative cost comparison is difficult because different topologies use different number of switches and different requirement on the switches and auxiliary circuit. But in general, due to the large power the charger is due dealt with, a FB converter is a more preferred choice for the main circuit.

4.4.1.8 Maximum charging ability

The previous analysis is based on the assumption that the load (output capacitor + battery) is regarded as an ideal voltage source. Without the filtering inductance, the primary and secondary voltage sources are connected in the switching process

Table 4.1   Device numbers for the different topologies.

	Forward	Flyback	Half-bridge	Full-bridge
Active devices	1	1	2	4
Diodes	2	1	2	4
Transformer	1	1	1	1
Inductor	1	Not needed	Not needed	Not needed
Capacitor	1	1	3	1
Snubber	1 set	1 set	0	0

through the leakage inductance of the transformer. In order to output sufficient power and guarantee a high efficiency, the leakage inductance of the transformer should not be too large, otherwise it will induce a large current peak in the semiconductors. In Figure 4.25, the solid line shows the maximum charging current of the battery, where the maximum repetitive turn-off current of the MOSFETs is 70 A, the turns ratio of the transformer is 1:1.5, and the input voltage is 400 V DC.

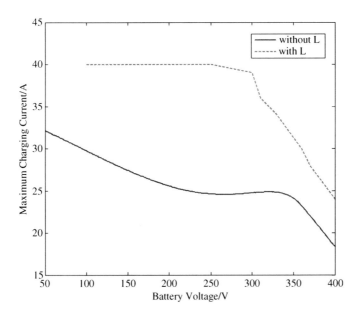

Figure 4.25   Maximum charging current at different battery voltage.

Adding a filtering inductance between the output capacitance and the secondary rectifier will not only smooth the charging current, but also help increase the maximum charging current, as shown by the dashed line of Figure 4.25. When this inductance is large enough, the current in the filtering inductor can be regarded as constant, therefore the peak primary-side current will be clamped to the charging current. The current ripple into the output capacitor is small. Therefore the maximum charging power is enhanced.

4.4.1.9 Power factor correction

The above analysis is based on the constant DC voltage of 400 V. For battery chargers that receive power from the AC grid, a front end rectifier is needed. In addition, a power factor correction (PFC) is needed to avoid injecting harmonics into the grid [23]. Figure 4.26 compares the grid current when the charger is

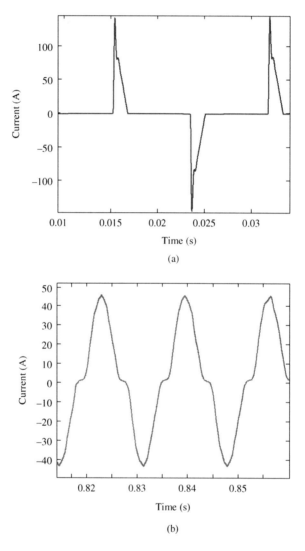

Figure 4.26   AC grid side current (a) without PFC and (b) with PFC.

equipped with and without the PFC. The impact on the grid is significantly reduced with a PFC.

In addition to correcting the power factor, the PFC will also boost the DC voltage to a higher constant value. Therefore, at the same output power, the switch current will be lower compared to chargers that do not contain the PFC, enhancing the safety and output capability of the charger.

Figure 4.27 shows the maximum charging current that the system can handle. Increasing the input voltage will benefit the output capability and therefore

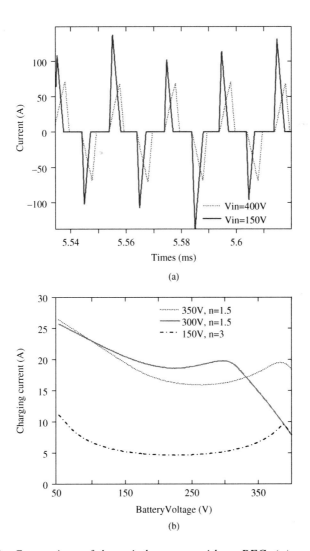

Figure 4.27   Comparison of the switch current without PFC: (a) comparison of switch current, (b) maximum charging current under different V_{in}.

shorten the charging time for a given circuit design. When V_{in} is increased from 150 to 400 V DC, the switch current is significantly decreased. As long as the voltage across the switches does not exceed the breakdown voltage, a higher DC-bus voltage will lead to a higher power capability of the charger. Figure 4.27b shows the maximum charging current that the system can deliver using the MOSFET whose maximum repetitive turn-off current is 70A.

4.4.2   Inductive charger

The above chargers all need an electrical contact with the outlet. This hard-wired electrical connection can provide a few caveats. For example, if the cable is pulled out from the electrical outlet (whether intentional or unintentional) when the battery is still being charged, then there could be a spark and potential damage or injury. Another example is where somebody (such as children) could get hurt if they happen to play with the cords etc. Charging the vehicle when it is raining could be potentially dangerous. Wear and tear of the plug could also be a source of danger. Hence inductive charging becomes an alternative. Figure 4.28 shows one of the topologies used in inductive chargers which can realize bidirectional power flow.

Inductive charging is a battery-charging technology which utilizes electromagnetic induction to provide a non-contact charging mode to transfer power. First, alternating current passes through the inverter unit and turns into high-frequency alternating current after the rectifier and inverter. The primary and secondary sides of the inductive coupler are tightly coupled through the electromagnetic field. Therefore, high-frequency alternating current flows through the secondary rectifier to charge the battery. The control signals, such as measurement of voltage, current, and temperature in the charging process, are collected by sensors and sent wirelessly to the controller unit located on the primary side of the coupler for feedback control or monitoring.

4.4.2.1   State-of-the-art inductive charging technology

In 1995, the US SAE Electric Vehicle Charging System Group, Japanese manufacturers including Toyota, Nissan, Honda, and DENSO, and the IEC/ISO together developed the inductive charging standard SAE J1773. In 1999, SAE J1773 was enacted and recommended as an international standard [25]. GM has already produced EVs adopting inductive charging in its EV1 program in the 1990s [24].

Since inductive charging does not involve the direct contact of electricity, it is a safe and convenient way to charge an EV battery. Inductive charging involves an isolation transformer whose secondary winding and core can be detached from the primary ones. The University of California at Berkeley has tested EVs on a specific path providing electric energy for the batteries. It can also drive vehicles at high speed without consuming energy in the batteries of the vehicles but by absorbing electric energy from the specific path. In 1999, Toyota produced an EV named RAV4LEV which contained an inductive charging system [26]. It was developed jointly by Toyota and GM from June 1998 with the long-term goal of

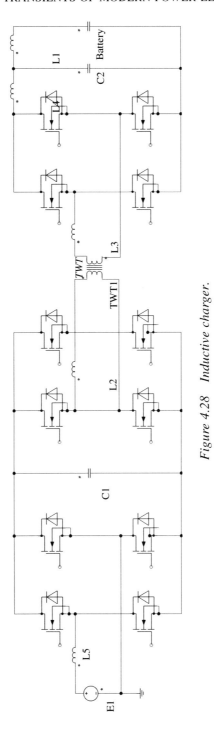

Figure 4.28   Inductive charger.

worldwide production. In New Zealand, EVs with inductive charging have been operating in Rotorua Geotherm National Park since 1999. In France, there are also large-gap inductive charging systems used for EVs [26].

4.4.2.2   Key technologies for inductive charging

The inductive coupler is the key component of an inductive charging system. Complying with the principle of electromagnetic induction, energy can be transferred from the primary to the secondary effectively and efficiently when both sides of the coupler are close to each other within a specific distance. There are three types of couplers on the market: the rotary, separable, and linear types [27]. The configurations of these three types of the couplers are shown in Figure 4.29.

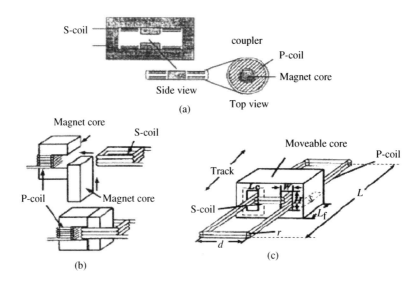

Figure 4.29   Configuration of the three types of couplers: (a) GM hughes, (b) separable type, and (c) linear type.

In applications such as EVs, robots, and so on, it is required that the coupler can work efficiently even with a large air gap and asymmetric structure. When the vehicle is charged, the air gap of the coupler cannot be guaranteed to be constant, resulting in a varying mutual inductance and thereby lower charging effectiveness and lower efficiency. In the process of using a variable gap coupler, the leakage inductance and magnetizing inductance will vary with the air gap, which is defined as the proximity effect. In real applications, the following issues need to be considered: (i) to enhance the assembly technology to make the electromagnetic coupling as tight as possible; and (ii) to make sure that the converter can adjust the control algorithms and generate the required power for any proximity.

Specifically, there are some general principles for the design of an inductive charging system:

1. **Soft-switching technologies for power devices [28, 29]:** in order to reduce the size of the coupler and increase the efficiency of electromagnetic coupling in the inductive charging system, a high-frequency inverter is necessary. With the PWM hard-switching technology, the switching loss of the semiconductor devices increases when the switching frequency increases. Further, in hard-switching mode, the high-order harmonic current of the inverter will increase the iron losses in the inductive coupler. The application of resonant soft-switching technology will reduce the switching losses of switches as well as the high-order harmonic components in the current.

2. **Wide-load-range operation [30]:** the charger should adapt to a wide load range. It is required to work properly whether the battery is completely discharged or nearly full. Also, it should be able to handle faults and implement the necessary protection properly and quickly, especially for short-circuit and open-circuit faults.

3. **Power factor correction:** in order to reduce electrical harmonics generated in the charging process, the PFC is very important for the system [31]. Based on the equivalent circuit of an inductive coupler and the designs above, Kutkut and Klontz [32] proposed a topology for a resonant power converter circuit (Figure 4.30) complying with the SAE J1773 standard.

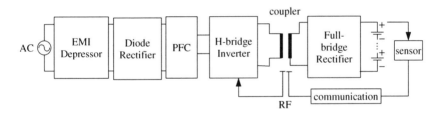

Figure 4.30   Recommended resonant converter circuit structure in SAE J1173.
© [1997] IEEE. Reprinted, with permission, from APEC'97.

In the design and application of different inductive charging systems, the main difference lies in the inverter. There are many types of resonant power inverters for inductive charging, such as full-bridge, half-bridge, boost, and so on. Some inductive charger designs do not include an EMI filter or PFC.

In the process of inductive charging, the control and feedback signals from the battery side are required to be fed back wirelessly. These signals include the terminal voltage of batteries, charging/discharging current, the temperature of batteries, SOC and SOH (State Of Health), and so on. Nowadays, there are two methods of wireless communication for inductive charging: RF communication and infrared communication. Through RF communication, the wireless communication and energy transmission can be integrated in one inductive coupler. Most inductive charging systems adopt RF for wireless communication. On the other hand, infrared communication has good anti-interference ability and can transmit information more reliably. The infrared communication could potentially become an international standard for inductive charging communication, avoiding the conflict among different frequency-band settings in different countries. However, it can only transmit information point to point, which is the main drawback compared to RF communication.

Because of the integration of high-power energy transmission and wireless transmission in one coupler, the high EMI between energy and information is inevitable. This is another important research aspect for inductive charging.

4.4.3   Wireless charger

Wireless charging involves the use of power and energy transfer over a much longer distance than in inductive charging. Hence, it is different from inductive charging, which involves a transformer with closely placed primary and secondary windings. Although inductive charging could eliminate direct electrical contact, it still needs a plug, a cable, and a physical connection of the inductive coupler. Wear and tear of the plug and cable could cause problems as well.

Wireless charging could eliminate the cable and plug altogether. In this scenario, a driver could pull the car over to a specially designed parking lot and the car battery would be automatically charged without pulling any cable or plug, as shown in Figure 4.31. Wireless charging provides the safest approach for EV battery charging.

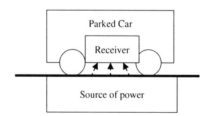

Figure 4.31   Wireless charging of a PHEV/EV on a parking floor.

There have been a few different experiments carried out on wireless energy transfer. The most promising technology is using electromagnetic resonance as shown in Figure 4.32. In this setup, there is a pair of antennas with one placed in the parking structure as the transmitter, and one inside the car as the receiver. The two antennas are designed to resonate at the controlled frequency. The limitations are the level of power transfer and efficiency due to the large air gap between the two antennas.

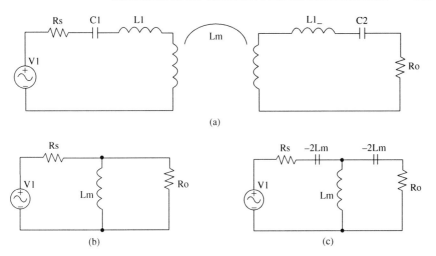

Figure 4.32   Circuits for electromagnetic resonance-based wireless charging: (a) circuit; (b) equivalent circuit at resonance frequency condition 1; and (c) equivalent circuit at resonance frequency condition 2.

In Figure 4.32, R_s is the internal resistance of the primary coil, R_o is the load resistance and the resistance of the secondary coil, L_1 is the leakage inductance of each coil, and L_m is the mutual inductance of the two coils. There are two ways to make the circuit resonant. If we design the circuit and select the frequency of the power supply V_1 such that $\omega L_1 - (1/\omega C) = -2\omega L_m$ and the total equivalent impedance is

$$Z = R_s - \frac{j\omega L_m R_o}{R_o - j\omega L_m}$$

Or we could design

$$\omega L_1 - \frac{1}{\omega C} = 0$$

and the total equivalent impedance is

$$Z = R_s + \frac{j\omega L_m R_o}{j\omega L_m + R_o}$$

then the circuit will be in resonance. However, since both the mutual inductance and leakage inductance change with the distance between the two coils, the frequency will have to be tuned based on the distance in real-world applications. Figure 4.32b,c shows the equivalent circuit during resonance of the two conditions. Figure 4.33 shows the simulation results of the circuit.

The first figure shows two resonant frequencies. The second figure shows that when the distance between the two coils increases, the two resonant frequencies get closer.

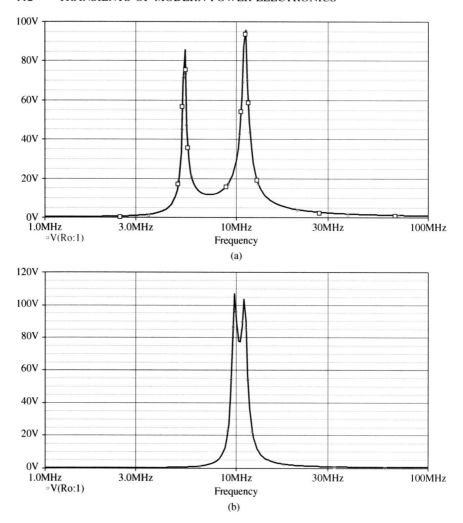

Figure 4.33   (a,b) Resonance frequency of the wireless charging circuit.

4.4.4   Optimization of a PHEV battery charger

Technically, an optimized charger should possess attributes such as high efficiency, low cost, low voltage ripple, small-size passive components (inductor
and capacitor), and so on. At the fixed output power, high efficiency and low
electrical stress of the semiconductors are important.

In theory, methods to decrease the peak current of the semiconductors are
either to increase the switching frequency or to increase the leakage inductance,
which will directly penalize the system efficiency and power capability. In this
section, the optimization of a PHEV charger is discussed in detail.

In order to precisely calculate the system efficiency, a model was implemented in MATLAB Simulink, including the transformer model and functional model of MOSFETs and diodes. The parameters of the MOSFETs are listed in Table 4.2.

Table 4.2   Key parameters of MOSFETs.

Name	Value
On-state resistance (R_{on})	65 mΩ
Rise time (t_r)	27 ns
Trailing time (t_f)	18 ns
Reverse recovery time (t_{rr})	<200 ns

When more than one design objective are in play, the design problem becomes a multi-objective optimization problem. For this design and optimization example, we choose the system efficiency and maximum device current as the optimization objective, subject to some boundary conditions:

$$\begin{array}{ll} \max & \eta \\ \min & I_{\max} \end{array}$$

$$\text{subject to} \quad \begin{array}{l} 1\,\text{kHz} \le f_s \le 200\,\text{kHz} \\ 10\,\mu\text{H} \le L_s \le 100\,\mu\text{H} \\ 50\,\text{A} < I_{\max} \le 90\,\text{A} \end{array} \qquad (4.47)$$

where η is the system efficiency, L_s is the equivalent leakage inductance of the transformer, and f_s is the switching frequency of the MOSFETs. In order to achieve higher power ratings for the charger, two MOSFETs are connected in parallel for each switch, with each of the switches handling a maximum of 70 A at 25 °C and 50 A at 75 °C. Considering the possible imbalance of current distribution in the parallel MOSFETs, the upper limit for the primary current peak $I_{\max}$ is set as 90 A. On the other hand, in order to fully utilize the MOSFETs, the lower boundary, 50 A, is set when the output power is 2.5 kW.

Equation 4.47 shows that obtaining the optimal parameters in this nonlinear system is a typical multi-objective problem dealing with more than one objective function. In this chapter, for demonstration purposes, the non-dominated sorting genetic algorithm (NSGA-II) developed by Deb and colleagues [33], one of the most efficient evolutionary algorithms, is used to optimize the system design. The computation steps for NSGA-II are conducted as shown in the flowchart of Figure 4.34.

The first step of this procedure is to initialize the size of population, generation number, numbers of objects, and variables. Here the variables are [n, f_s, L_s]. The individual is generated randomly within the boundary mentioned above. After initialization, those individuals are regarded as the parents. Further effort is made to generate the offspring of these parents by the binary crossover operator, mutation,

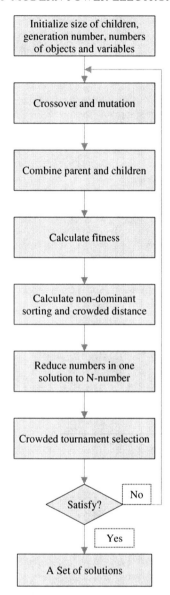

Figure 4.34   Flowchart of the genetic algorithm.

non-dominant sorting, and tournament selection. More details of this algorithm can be found in [34–36]. The book by Mi *et al.* [37] also contains further details covering not only algorithms, but also many of the points discussed above.

In exploring the system capability while minimizing I_{max} and maximizing the efficiency of the system, some constraints are inevitable. For example, the

MOSFET current should not be too large and endanger the system or too low and underutilize the MOSFET capability. The switching frequency should not be too high and cause extra heat loss or too low and not make the best use of the MOSFET's advantages.

Without limiting the current peak, the results converging after a 10-generation optimization (Figure 4.35) are

$$L_s = 70\,\mu\text{H}, \; f_s = 44\,\text{kHz}, \; n = 2.8, \; \text{efficiency} = 85.14\%, \; I_{\text{max}} = 48\,\text{A}$$

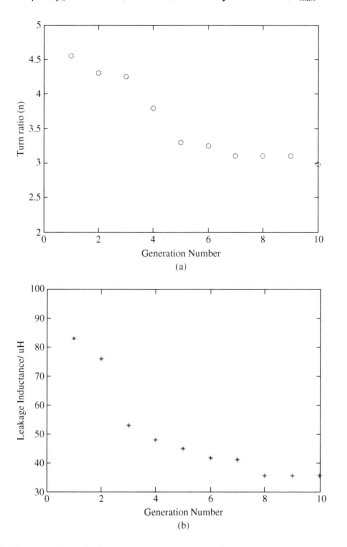

Figure 4.35   Results of the optimization procedure: (a) optimization of turns ratio, (b) optimization of the leakage inductance, and (c) optimization of the switching frequency.

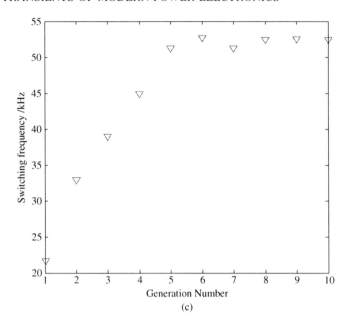

(c)

Figure 4.35   (continued)

From these results, a peak current of 48 A is far less than the current capability of the system. Minimizing I_{max} without any constraints leads directly to a leakage inductance of 70 μH, which is too large to be integrated inside the transformer. The efficiency is also low.

The randomly selected parents and the generated children are taken into the model set in Simulink to execute the simulation. The efficiency of the system can be calculated by the Simulink model and maximum current can also be obtained when the system reaches steady state operation. Here the optimal solution for this DC–DC converter converging after 10-generation operation is

$$n = 2.98, L_s = 35.5 \, \mu H, f_s = 52.6 \, kHz,$$

$$\text{efficiency} = 88.3\%, I_{max} = 59.6 \, A$$

4.4.5   Bidirectional charger and control

With bidirectional power transfer capability, the energy stored in a PHEV battery can be sent back to the grid during peak demand hours for peak shaving of the AC electric grid, or to supply power to the home and office during a power outage.

Isolation can be achieved by using high-frequency transformers in the DC–DC stage at high-frequency level, as shown in Figure 4.36. The isolation can also be achieved by using a transformer at the grid frequency level as shown in Figure 4.37. It is difficult to add a PFC stage to the bidirectional chargers but

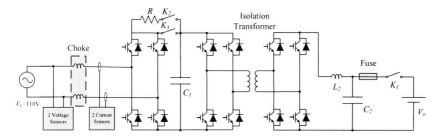

Figure 4.36   Isolation using a high-frequency transformer.

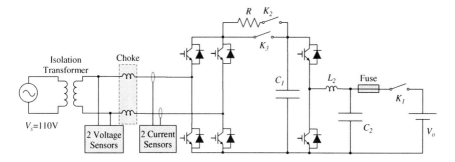

Figure 4.37   Isolation at the grid level with a line-frequency transformer.

the AC side current can be controlled using the grid side inverter to limit the harmonics and improve power factor.

In this section, we will discuss the grid side-isolated bidirectional charger as shown in Figure 4.37 but extended to a three-phase application. The three inductors between the three-phase AC grid and FB converter are necessary to boost the DC-bus voltage to charge the battery which is typically at a higher voltage (200–400 V) than the AC side voltage (208 V).

In charge mode, the DC-bus voltage must be higher than the battery voltage. The upper switch of the DC–DC converter is subjected to PWM control and the lower switch is kept off. The circuit becomes a buck circuit. Through controlling the PWM duty ratio of the upper switch, the charging current or voltage can be controlled.

In discharge mode, the upper switch of the DC–DC converter is kept off and the lower switch is under the PWM mode. The whole circuit becomes a backward boost circuit. Simultaneously, the phase angle of the power grid is obtained, in order to adjust the power factor of the output power from the inverter.

In the following discussion, the controllers are realized through dSPACE and MATLAB/Simulink.

According to the hardware configuration, two separate control algorithms are needed for system-level realization, namely, a PWM rectifier control algorithm and a DC–DC converter control algorithm.

Due to the strong coupling in the three-phase power, a decoupled control method should be applied similar to the field-oriented control strategy used in a three-phase induction motor. It could decouple three-phase AC variables and transfer to two separate DC variables. In the suggested control algorithm shown in Figure 4.38, the rectifier controller is realized based on the direct current control (DCC) algorithm. In this way, the DC-bus voltage can be regulated effectively, a near-unity power factor can be reached, and the total harmonic distortion (THD) can be decreased dramatically. As for the DC–DC converter, a typical proportional–integral (PI) and bang–bang controller can be implemented.

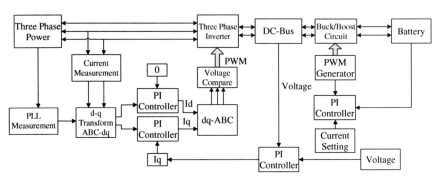

Figure 4.38   Control algorithm of the grid side isolated bidirectional charger.

In order to apply the DCC algorithm, the following five steps are needed and some calculations are necessary to decouple the system. First, the magnitude and phase angle of the three-phase alternating current need to be measured; second, a d–q transformation is used to get the decoupled variable; third, the two decoupled variables are controlled using an intelligent controller; and finally the inverse d–q transformation is applied based on the controller output and to generate the control signals.

The measurement of magnitude and angle of three-phase voltage is based on the phase-locked loop (PLL) principle. The PLL is a feedback control system that automatically adjusts the phase of a locally generated signal to match the phase of an input signal. The signal's frequency and angle signal can be directly obtained.

The equivalent circuit for the three-phase inverter is shown in Figure 4.39. The transfer function of the system is analyzed below.

The on/off function of each bridge is defined as

$$S_k = \begin{cases} 1 & \text{the upper switch turns on, the lower switch turns off} \\ 0 & \text{the upper switch turns off, the lower switch turns on} \end{cases}$$

$$k = a, b, c \tag{4.48}$$

The system function is described as

$$Z\dot{X} = AX + BE \tag{4.49}$$

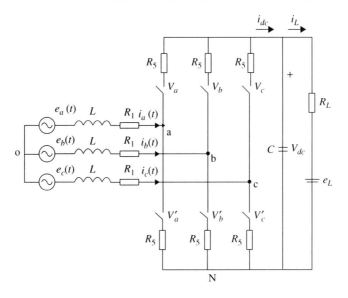

Figure 4.39   DC PWM rectifier.

where $X = [i_a, i_b, i_c, v_{dc}]^T$ and i_a, i_b, i_c denote the current flow through phases A, B, and C. The DC-bus voltage is denoted by v_{dc}. Then

$$
A = \begin{bmatrix}
-R & 0 & 0 & -\left(s_a - \dfrac{1}{3}\displaystyle\sum_{k=a,b,c} s_k\right) \\[2.5em]
0 & -R & 0 & -\left(s_b - \dfrac{1}{3}\displaystyle\sum_{k=a,b,c} s_k\right) \\[2.5em]
0 & 0 & -R & -\left(s_c - \dfrac{1}{3}\displaystyle\sum_{k=a,b,c} s_k\right) \\[2.5em]
s_a & s_b & s_c & -1/R_L
\end{bmatrix}
\tag{4.50}
$$

where R is the inductor's internal resistance and R_L is the load resistance, and

$$
Z = \begin{bmatrix}
L & 0 & 0 & 0 \\
0 & L & 0 & 0 \\
0 & 0 & L & 0 \\
0 & 0 & 0 & C
\end{bmatrix}
\tag{4.51}
$$

$$B = \begin{bmatrix} 1 & 0 & 0 & 0 \\ 0 & 1 & 0 & 0 \\ 0 & 0 & 1 & 0 \\ 0 & 0 & 0 & \dfrac{1}{R_L} \end{bmatrix} \tag{4.52}$$

$$E = [e_a, e_b, e_c, e_L]^T \tag{4.53}$$

$$\dot{X} = \left[\frac{di_a}{dt}, \frac{di_b}{dt}, \frac{di_c}{dt}, \frac{dv_{dc}}{dt} \right]^T \tag{4.54}$$

It can be seen from Equations 4.48–4.54 that there is strong coupling between several variables of the three-phase system, increasing the complexity of the control algorithm. The control inputs are the switch status of the three bridges, and the output is the DC-bus voltage. Further, harmonics and power factor become other issues needing consideration in order to decrease pollution to the power grid.

As in control of the induction motor, the d–q axis transform is an effective way to solve the problem. The d–q transformation formula is

$$i_d = \tfrac{2}{3}[i_a \sin(wt) + i_b \sin(wt - \tfrac{2}{3}\pi) + i_c \sin(wt + \tfrac{2}{3}\pi)]$$
$$i_q = \tfrac{2}{3}[i_a \cos(wt) + i_b \cos(wt - \tfrac{2}{3}\pi) + i_c \cos(wt + \tfrac{2}{3}\pi)] \tag{4.55}$$
$$i_0 = \tfrac{2}{3}[i_a + i_b + i_c]$$

where wt is the angle and i_d, i_q denote the d-axis and q-axis current.

After transformation, Equation 4.49 can be expressed as

$$C\frac{dv_{dc}}{dt} = \frac{3}{2}(i_q s_q + i_d s_d) - i_L$$
$$L\frac{di_q}{dt} + wLi_d + Ri_q = e_q - v_{dc}s_q \tag{4.56}$$
$$L\frac{di_d}{dt} + wLi_q + Ri_d = e_d - v_{dc}s_d$$

In this d–q system, each phase's current is decoupled. By controlling i_d and i_q, the DC-bus voltage can be controlled directly.

As shown in Figure 4.39, to control i_d and i_q separately, two PI controllers are used. In order to achieve a unity power factor, the control objective of i_d is zero, and that of i_q is determined by the DC-bus voltage feedback.

After obtaining the desired i_d and i_q values, the inverse transform from the d–q coordinate to the abc coordinate is applied and the demanded i_a, i_b, i_c can be obtained. Thus,

$$i_a = i_d \sin(wt) + i_q \cos(wt)$$
$$i_b = i_d \sin(wt - \tfrac{2}{3}\pi) + i_q \cos(wt - \tfrac{2}{3}\pi) \tag{4.57}$$
$$i_c = i_d \sin(wt + \tfrac{2}{3}\pi) + i_q \cos(wt + \tfrac{2}{3}\pi)$$

For the DC–DC converter, the system is realized through another PI controller. The control target is to regulate the battery current or voltage to reach the desired value quickly and without too much overshoot. The control variable is the duty ratio of the PWM waveform for the power switches. The PI controller uses the battery current and voltage as feedback. Both constant current charging and constant voltage charging can be realized. Pulsed charge is also possible by adjusting the control algorithms. In the experiments presented below, we first use constant current charging until the battery voltage reaches a preset value, and then we switch to constant voltage charging. Due to noise and small fluctuations, it is very difficult to control the PWM duty ratio directly based on voltage feedback under the constant voltage charging mode. Based on this characteristic, the system adopts a bang–bang control, which uses voltage feedback to change the current setting, operating in the current control mode, as shown in Figure 4.40. With this implementation, constant voltage charging is realized indirectly. In Figure 4.40, a_1 and a_2 are constants that can be determined by analyzing the system response time and inertia.

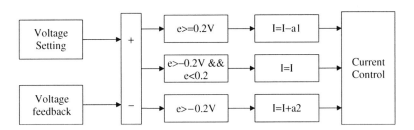

Figure 4.40   Constant voltage charging control strategy.

We use simulation to validate the functions of the charger, define the parameters of the controller, and observe the performance of the system. The whole system simulation is performed in MATLAB/Simulink. A three-phase inverter, a DC–DC converter circuit, and a battery model from the SimPowerSystems library located in Simulink are used for the simulation setup. The main elements' parameters are listed in Table 4.3 using data from the hardware setup. The system is designed for a battery pack with a rated voltage of 350 V, rated capacity of 40 Ah, and recommended maximum charging current of 25 A.

4.4.5.1 Simulation of constant current charging

In this simulation, the charging current is set to 15 A and the DC-bus voltage is set to 350 V; the DC-bus voltage and battery current waveforms are shown in Figure 4.41. Figure 4.41a shows the DC-bus voltage is maintained at the set point with very small ripple. Figure 4.41b shows the charging current reaching the set point without overshoot. The current ripple at steady state is within 0.5 A.

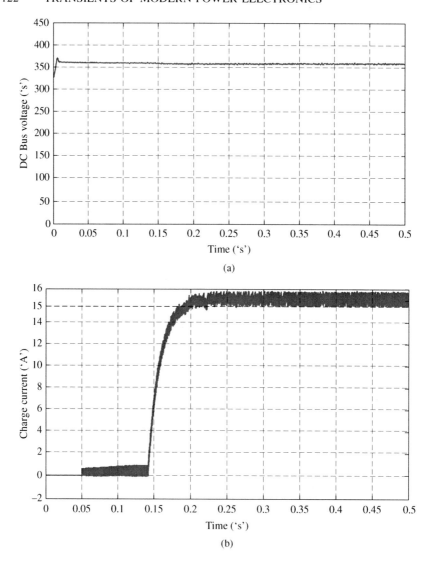

Figure 4.41   Constant current charging simulated using MATLAB/Simulink: (a) DC-bus voltage when charging the battery at 15 A rate; (b) battery current waveform under charging status.

4.4.5.2   Constant current discharging simulation

The discharging current is set to 15 A. The simulated current waveform during constant discharge is shown in Figure 4.42. The current can track the set point without overshooting.

Table 4.3   Simulation and hardware parameters.

Unit	Value
Three-phase inductance	Resistance: 0.2 Ω Inductance: 1.8 mH
Three-phase AC voltage	110 V, Y-type
Universal bridge	Snubber resistance R_s: 1×10^5 Ω R_{on}: 1×10^{-3} Ω Fall time: 1×10^{-6} s Trailing time: 2×10^{-6} s
Filtering capacitance	4950×10^{-6} F
DC–DC converter inductance	Resistance: 0.2 Ω Inductance: 3.3 mH
Battery	Type: lithium-ion battery Rated voltage: 300 V Capacity: 40 Ah

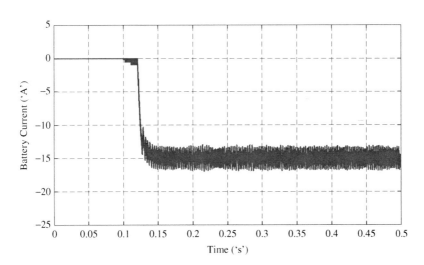

Figure 4.42   Battery current waveform when discharging the battery.

4.4.5.3   Experiment

The hardware setup is shown in Figure 4.43. The hardware includes dSPACE DS1104, a three-phase PWM rectifier with a maximum operating voltage of 600 V, maximum current 200 A, and maximum operating frequency 20 kHz.

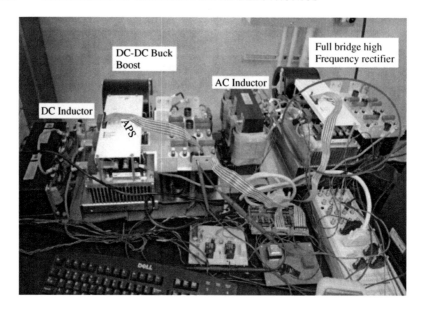

Figure 4.43   The hardware system.

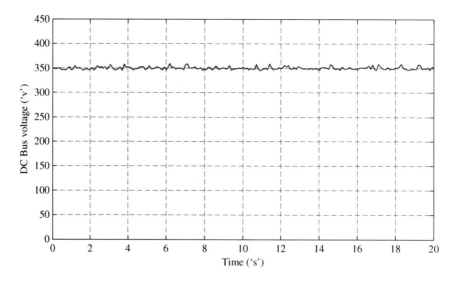

Figure 4.44   DC-bus voltage when charging the battery at 10 A rate.

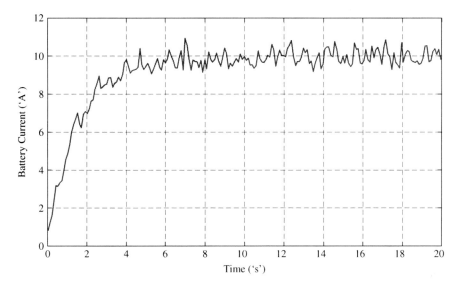

Figure 4.45   Battery current when charging.

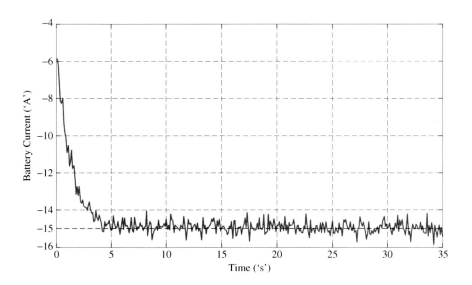

Figure 4.46   Battery current waveform when discharging.

The charge current is set to 10 A and the DC-bus voltage is set to 350 V. The experimental results are shown in Figures 4.44 and 4.45. It can be seen that the current follows the command very well. The ripple is less than 0.5 A.

When the battery is delivering power to the grid, it is discharged. The discharge current waveform is shown in Figure 4.46. The response time is less than 5 seconds. There is no overshoot in the response.

References

1. Yu-shan, L., Qing-liang, Z., Cheng-long, W., and Liang, W. (2009) Research on control strategy for regenerative braking of a plug-in hybrid electric city public bus. International Conference on Intelligent Computation Technology and Automation, pp. 842–845.

2. Lu, R., Wu, G., Ma, R., and Zhu, C. (2008) Model based state of charge estimation method for ultra-capacitor. IEEE Vehicle Power and Propulsion Conference, pp. 1–5.

3. Ellis, M.W., Von Spakovsky, M.R., and Nelson, D.J. (2001) Fuel cell systems: efficient, flexible energy conversion for the 21st century. *Proceedings of the IEEE*, **89** (12), 1808–1818.

4. Zhang, X., Mi, C., Masrur, A., and Daniszewski, D. (2008) Wavelet based power management of hybrid electric vehicles with multiple onboard power sources. *Journal of Power Sources*, **185** (2), 1533–1543.

5. Sun, L., Chan, C.C., Liang, R., and Wang, Q. (2008) State-of-art of energy system for new energy vehicles. Vehicle Power and Propulsion Conference, pp. 1–8.

6. Yoo, H., Sul, S.-K., Park, Y., and Jeong, J. (2008) System integration and power-flow management for a series hybrid electric vehicle using supercapacitors and batteries. *IEEE Transactions on Industry Applications*, **44** (1), 108–114.

7. Li, X. and Williamson, S.S. (2007) Comparative investigation of series and parallel hybrid electric vehicle (HEV) efficiencies based on comprehensive parametric analysis. IEEE Vehicle Power and Propulsion Conference, pp. 499–505.

8. Mekhiche, M., Nichols, S., Kirtley, J.L. *et al.* (2001) High-speed, high-power density PMSM drive for fuel cell powered HEV application. IEEE International Electric Machines and Drives Conference, pp. 658–663.

9. Wu, T., Chi, Y.-L., Guo, Y., and Xu, C. (2009) Simulation of FOC vector control of induction motor based on LabVIEW. International Conference on Information Engineering and Computer Science, pp. 1–3.

10. Lanhong, Z., Yuwen, H., and Wenxin, H. (2005) Research on DTC control strategy of induction starter/generator system. Proceedings of the Eighth International Conference on Electrical Machines and Systems, pp. 1528–1533.

11. Morimoto, M., Sumito, K., Sato, S. *et al.* (1991) High efficiency, unity power factor VVVF drive system of an induction motor. *IEEE Transactions on Power Electronics*, **6** (3), 498–503.

12. Hua, B., Zhengming, Z., Liqiang, Y., and Bing, L. (2006) A high voltage and high power adjustable speed drive system using the integrated LC and step-up transforming filter. *IEEE Transactions on Power Electronics*, **21** (5), 1336–1346.

13. Zhao, Z., Bai, H., Meng, S. *et al.* (2003) An effective method for neutral point balance of AC inverters. International Conference on Electrical Machines and Systems, pp. 372–374.

14. Bai, H., Zhao, Z., and Mi, C. (2009) Framework and research methodology of short-timescale pulsed power phenomena in high voltage and high power converters. *IEEE Transactions on Industrial Electronics*, **56** (3), 805–816.

15. Sung-Sae, L., Seong-Wook, C., and Gun-Woo, M. (2007) High-efficiency active-clamp forward converter with transient current build-up (TCB) ZVS technique. *IEEE Transactions on Industrial Electronics*, **54** (1), 310–318.

16. Wang, C.-M. (2008) A novel ZCS-PWM flyback converter with a simple ZCS-PWM commutation cell. *IEEE Transactions on Industrial Electronics*, **55** (2), 749–757.

17. Fathy, K., Morimoto, K., Doi, T. *et al.* (2006) A new soft-switching PWM half-bridge DC-DC converter with high and low side DC rail active edge resonant snubbers. IEEE Power Electronics Specialists Conference, pp. 1–7.

18. Zhang, J., Xie, X., Wu, X. *et al.* (2006) A novel zero-current-transition full-bridge DC/DC converter. *IEEE Transactions on Power Electronics*, **21** (2), 354–360.

19. Mi, C., Bai, H., Wang, C., and Gargies, S. (2008) The operation, design, and control of dual H-bridge based isolated bidirectional DC-DC converter. *IET Power Electronics*, **1** (3), 176–187.

20. Wang, C.-M., Su, J.-H., and Yang, C.-H. (2004) Improved ZCS-PWM commutation cell for IGBTs application. *IEEE Transactions on Aerospace and Electronic Systems*, **40** (3), 879–888.

21. Mao, H., Abu-Qahouq, J., Luo, S., and Batarseh, I. (2004) Zero-voltage-switching half-bridge DC-DC converter with modified PWM control method. *IEEE Transactions on Power Electronics*, **19** (4), 947–958.

22. Wu, X., Xie, X., Zhao, C. *et al.* (2008) Low voltage and current stress ZVZCS full bridge DC–DC converter using center tapped rectifier reset. *IEEE Transactions on Industrial Electronics*, **55** (3), 1470–1477.

23. Huber, L., Gang, L., and Jovanovic, M.M. (2010) Design-oriented analysis and performance evaluation of buck PFC front end. *IEEE Transactions on Power Electronics*, **25** (1), 85–94.

24. SAE International (1999) SAE Electric Vehicle Inductively Coupled Charging, J173_200905, November 1.

25. GM ATV (1998) WM7200 Inductive Charger Owner's Manual, http://www.evchargernews.com/.

26. Zhi-yu, L., Dong, D., and Guo-gang, Q. (2004) Development and application of inductive charging technology. *Power Electronics*, **38** (3), 92–94.

27. Hiraga, Y., Hirai, J., Kaku, Y. *et al.* (1994) Decentralized control of machines with the use of inductive transmission of power and signal. Proceedings of the IEEE/IAS Annual Meeting, Vol. 2, pp. 875–881.

28. Zhu, L. (2006) A novel soft-commutating isolated boost full-bridge ZVS-PWM DC-DC converter for bidirectional high power application. *IEEE Transactions on Power Electronics*, **21**, 422–429.

29. Fang, H.L. and Peng, Z.Z. (2004) Modelling of a new ZVS bi-directional DC-DC converter. *IEEE Transactions on Aerospace and Electronic Systems*, **40** (1), 272–283.

30. Hayes, J.G., Egan, M.G., Murphy, J.M.D. *et al.* (1999) Wide-load-range resonant converter supplying the SAE J-1773 electric vehicle inductive charging interface. *IEEE Transactions on Industry Applications*, **35** (4), 884–895.

31. Qiao, C. and Smedley, K.M. (2001) A topology survey of single-stage power factor corrector with a boost type input-current-shaper. *IEEE Transactions on Power Electronics*, **16** (3), 360–368.

32. Kutkut, N.H. and Klontz, K.W. (1997) Design considerations for power converters supplying the SAE J-1773 electric vehicle inductive coupler. Proceedings of the IEEE Applied Power Electronics Conference and Exposition, Vol. 2, pp. 841–847.

33. Srinivas, N. and Deb, K. (1994) Multiobjective optimization using nondominated sorting in genetic algorithms. *Evolutionary Computation*, **2** (3), 221–248.

34. Malyna, D.V., Duarte, J.L., Hendrix, M.A.M., and van Horck, F.B.M. (2007) Optimization of combined thermal and electrical behavior of power converters using multi-objective genetic algorithms. European Conference on Power Electronics and Applications, pp. 1–10.

35. Huang, B., Wang, Z., and Xu, Y. (2006) Multi-objective genetic algorithm for hybrid electric vehicle parameter optimization. IEEE/RSJ International Conference on Intelligent Robots and Systems, pp. 5177–5182.

36. Zhang, Z. and Cuk, S. (2000) A high efficiency 500 W step-up Cuk converter. International Power Electronics and Motion Control Conference, pp. 909–914.

37. Mi, C., Masrur, A., and Gao, D. (2011) *Hybrid Electric Vehicles*, John Wiley & Sons, Ltd, Chichester.

5

Power electronics in alternative energy and advanced power systems

5.1 Typical alternative energy systems

The dominant energy sources today are primarily fossil-based energy sources, that is, petroleum, natural gas, coal, and, in some cases, nuclear energy. With the development of economies and growth of populations, it is expected that in the near future the fossil-based sources will be depleted very quickly. Alternative energy sources, such as solar energy, wind energy, wave/tidal energy, and biomass-based energy are expected to grow at a rapid pace and eventually replace the primary fossil energy in the near future.

Compared to the traditional fossil energy sources, these alternative sources have less carbon emissions and are sustainable. However, with the limitations from environmental and climate conditions, most alternative energy sources, such as solar and wind, are not stable sources. Power generated by a photovoltaic (PV) system is strongly related to solar intensity, and the energy provided by a wind system is directly determined by wind speed. These different forms of energy must be controlled to obtain a stable and high-quality power output through power electronics.

In the next section, the role of power electronics in solar and wind energy systems will be addressed. The dynamics and transients of such systems will be discussed. When solar or wind energy systems are interfaced with the grid, the requirements are even more stringent. First, the power capability of the alternative energy system should be maximized to achieve the minimal payback period,

Transients of Modern Power Electronics, First Edition. Hua Bai and Chris Mi.
© 2011 John Wiley & Sons, Ltd. Published 2011 by John Wiley & Sons, Ltd.

which directly determines whether the alternative energy is economically viable. Second, the operation of the alternative energy system should not impact the quality of the electric grid it is connected to. This is of particular importance in the transient processes.

5.2   Transients in alternative energy systems

A typical solar energy system is shown in Figure 5.1 [1]. In this system, a PV array is the power source. It absorbs solar energy and transforms it to electrical energy, and is followed by a DC–DC converter. If the load has AC characteristics, for example, the electric grid or an induction motor, an extra DC–AC inverter is required. Any excessive solar energy will be stored in the local battery.

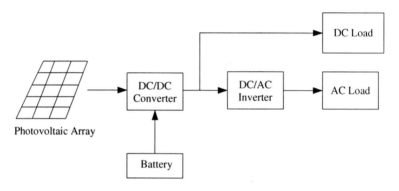

Figure 5.1   Photovoltaic electrical generating system.

As an example, when the DC loads in Figure 5.1 are light bulbs and the AC load is a water pump, the system appears as a stand-alone power system. During the daytime, the PV array generates energy to drive the pump. Any excessive energy will be stored in the battery. At night, the PV array stops working and the battery delivers the energy to supply light bulbs. Figure 5.2 shows the direction of power flow of the system at night and during the day.

5.2.1   Dynamic process 1: MPPT control in the solar energy system

The typical electrical characteristics of a PV panel, that is, the volt–ampere curve, are shown in Figure 5.3 [2]. These curves always change with solar intensity. With the same solar intensity, the PV panel will behave as a constant current source under light load and a quasi-constant voltage source under heavy load. In Figure 5.3b, the output power will reach the peak M, defined as the maximum power point. Optimal control should excavate the potential of the PV array, that

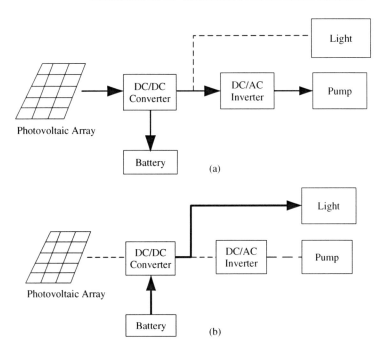

Figure 5.2   Energy flow inside the pumping and lighting system: (a) daytime operations and (b) night operations.

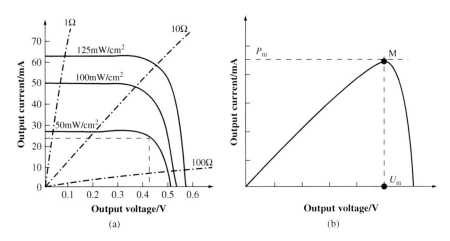

Figure 5.3   Characteristics of the photovoltaic battery: (a) volt−ampere curve of photovoltaic battery and (b) maximum output power. © [2006] China Science Publications, reproduced from Zhengming Zhao, Jianzheng Liu, Liqiang Li, Solar Photovoltaic Power Generation and Applications.

is, the system is controlled to deliver the maximum output power regardless of the change in solar intensity. This control is called maximum power point tracking (MPPT) [3], which is also the most typical transient process in the PV energy system.

Figure 5.4 shows how solar intensity can affect operation of the control. With solar intensity 1, the maximum power point is located at A if the load curve is load line 1. When the solar intensity is increased to 2, the maximum power point will be shifted to B'. However if the load line is still maintained as load line 1, the power point will shift from A to A' which is not the maximum power point of the PV system. In this case, if the load line is adjusted from line 1 to line 2, the maximum power (point B') can be extracted from the system. This is the conceptual procedure of MPPT.

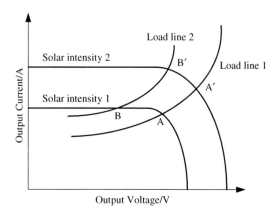

Figure 5.4   Trace of power tacking.

The output power of the PV array is always determined by the product of the output voltage and current, that is, $P = UI$ for the classic MPPT control, when the power reaches the peak

$$\frac{dP}{dU} = \frac{d(UI)}{dU} = U\frac{dI}{dU} + I\frac{dU}{dU} = 0 \tag{5.1}$$

Therefore

$$U\,dI = -I\,dU \tag{5.2}$$

Discrete dI and dU at time k will lead to

$$dI(k) = I(k+1) - I(k)$$
$$dU(k) = U(k+1) - U(k) \tag{5.3}$$

After substituting Equation 5.3 into Equation 5.2, the system will reach the maximum power point when

$$U(k)I(k+1) + I(k)U(k+1) - 2U(k)I(k) = 0 \tag{5.4}$$

Modern MPPT control has been tremendously extended, such as, perturb and observe algorithms [4], incremental conductance control [5], fuzzy logic control [6], and so on. However, the most commonly used control in PV systems is still the classic MPPT control.

5.2.2 Dynamic processes in the grid-tied system

Alternative energy systems are classified into two major types based on their relationship with the electric grid: stand-alone systems and grid-connected systems. In addition to many similarities in topology, they differ from each other in terms of control algorithms. Mostly a stand-alone system is used in off-grid applications with battery storage. One of the significant disadvantages of the stand-alone system is that a power insufficiency or excess will frequently emerge. When PV panels are shaded and the local battery is exhausted in the stand-alone PV energy system shown in Figure 5.1, the power capability of the system will decrease or disappear – hence no sufficient power can be supplied to meet consumer demands. When the local battery storage is full, excessive solar energy coming from the PV panel cannot be further stored, therefore the system is underutilized.

For a grid-connected system, the excessive power from the PV panel can be sold to the grid at peak time, and the grid power can support the customers as a backup or during shortages. Therefore, the grid-connected renewable energy system is much more attractive. A grid-connected solar energy-based three-phase voltage source inverter is shown as Figure 5.5.

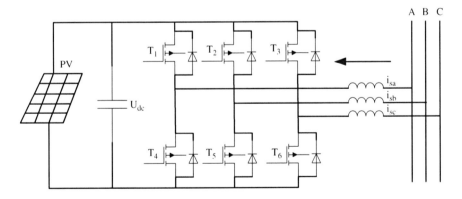

Figure 5.5   Voltage source in grid-connected system.

With increasing complexity of the system, conflicts in its dynamic processes will emerge. Those conflicts will be amplified when the alternative energy enters the high-power domain. For the solar energy system, typical examples include the sampling delay, shading effect, and islanding effect.

5.2.2.1 Sampling delay

As we mentioned above, the PV system will generate power to the grid during the daytime, acting as a power plant. At night-time, the PV system could behave as a reactive-power compensator [7]. The operational modes and equivalent circuits of Figure 5.5 are shown in Figure 5.6, where U_s is the grid phase voltage, U_i is the phase voltage of the inverter bridge, I is the inverter current, and $R + jX$ is the equivalent impedance of the inductor between the grid and inverter.

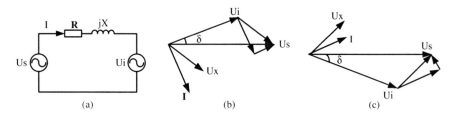

Figure 5.6   Operation modes of the voltage source in grid-connected system: (a) equivalent circuit, (b) when grid voltage lags behind the inverter's voltage, and (c) when grid voltage leads the inverter's voltage.

When U_s lags behind U_i, U_s will lead I, shown in Figure 5.6b. This means that the grid will generate the active power to the inverter. Otherwise, the active power will flow from the inverter to the grid. If δ, the angle between U_s and U_i, is close to zero, the inverter will behave as a reactive-power compensator.

From Figure 5.6a

$$\dot{U}_s = \dot{I}(R + jX) + \dot{U}_i \qquad (5.5)$$

If the resistance R is neglected, then

$$\dot{U}_s = \dot{I}\,jX + \dot{U}_i \qquad (5.6)$$

Therefore the grid current is

$$\dot{I} = \frac{\dot{U}_s - \dot{U}_i}{jX} = \frac{(U_s - U_i \sin \delta) + jU_i \cos \delta}{jX} \qquad (5.7)$$

The power can be defined as

$$S = \dot{U}_s\,\dot{I}^* = \frac{U_s U_i \sin \delta}{X} + j\frac{U_s(U_s - U_i \cos \delta)}{X} = P + jQ \qquad (5.8)$$

That is,

$$P = \frac{U_s U_i \sin \delta}{X}$$

$$Q = \frac{U_s (U_s - U_i \cos \delta)}{X}$$

(5.9)

If $\delta > 0$, that is, U_s leads U_i, the active power P will flow from the grid to the inverter. When δ is small enough, then $\sin \delta \approx \delta$ and $\cos \delta \approx 1$. Therefore the reactive power is determined by the amplitude of U_s and U_i. If $U_s > U_i$ then $Q > 0$. This means that the grid will generate the inductive reactive power from the grid, otherwise it is capacitive reactive power.

Macroscopically the above control strategy assumes that the current and voltage can be sampled and implemented without any time delay. However, not only does the time delay exist in the sampling of the current and voltage, but also the calculation of the amplitude and angle of U_i based on Equation 5.9 consumes computation time in the microcontroller. In the digital system, the calculated U_i in the present switching period can only be carried out in the next switching period, that is, the time delay is $1/f_s$. Here f_s is the switching frequency. We can call it a dead zone of the control algorithm, in which the voltage vector of the grid is still rotating. Therefore the calculated error of the angle caused by the time delay is

$$\Delta \delta = 2\pi \frac{T_s}{T_g}$$

(5.10)

where T_s is the switching period and T_g is the grid period. For example, if the grid frequency is 60 Hz and the switching frequency is 10 kHz, the angle error is $60/10000 \times 360° = 2.16°$. In a high-power and high-voltage system, the switching frequency is not allowed to be that high [8]. Therefore the angle error tends to be amplified. For example, if $f_s = 1$ kHz, the angle error will be $60/1000 \times 360° = 21.6°$, which cannot be neglected in the control system.

Figure 5.7a shows the case when the grid voltage leads the inverter voltage. Due to the time delay of one switching period, when U_i is actually imposed on the grid, the grid voltage vector has already rotated clockwise for an extra angle $\Delta \delta$, shown as U_{s2} and marked with a dotted line. Hence the real angle between U_s and U_i is enlarged to $\delta + \Delta \delta$. In Figure 5.7b where U_i is leading U_s, the real angle is also $\delta + \Delta \delta$. Note that $\delta < 0$. When $\Delta \delta$ is large enough, U_{s2} can even lead U_i again, which transforms the capacitive reactive power into inductive power.

The above calculated error caused by the sampling delay needs to be compensated when the switching frequency is low. The compensated value is $\Delta \delta$, shown in Equation 5.10. Whether the grid voltage leads or lags the inverter voltage, the angle of the inverter voltage φ should always be increased with an additional angle $\Delta \delta$, that is, $\varphi = \delta + \Delta \delta$.

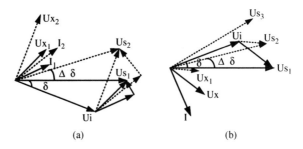

Figure 5.7   Influence of the error of the sampling delay: (a) U_s *leads* U_i *and (b)* U_s *lags* U_i.

5.2.2.2   Solar shading or diversity of the PV panels

When power demand increases, PV panels need to be connected in series and parallel as shown in Figure 5.8a. For series-connected solar panels, besides the potentially hazardous high voltage, this topology has two other shortcomings: (i) the sampling delay will become dominant when the switching frequency drops, as shown in Figure 5.7; and (ii) the MPPT control targets the whole panel string instead of maximizing the power capability of each PV panel. This will reduce the system power capability when solar panels have diversity; for example, some panels are shaded or suddenly fail. If several PV panels fail, or if the converter side encounters technical problems, the whole system will be halted.

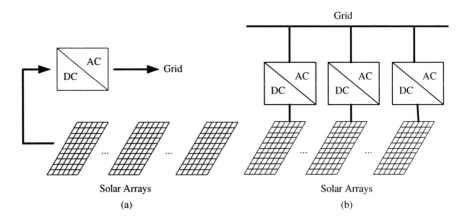

Figure 5.8   Comparison of string inverter and micro-inverters [9]: (a) single-string inverter and (b) micro-inverters.

The above conflicts can be mitigated through micro-inverters. The micro-inverter shown in Figure 5.8b is a feasible topology to maximize the power capability of each individual panel, where each panel is connected to one inverter, respectively. The micro-inverter successfully alleviates conflict in the high-power domain by using multiple low-power modules. If one PV panel or one inverter fails, other microsystems can still provide energy to the grid.

The power capability of one micro-inverter is around 200 W. Assuming the average renewable power supplied to a house is 2 kW, 10 micro-inverters will therefore be needed. More than 500 micro-inverters are needed if a 100 kW solar energy station is built.

5.2.2.3 Islanding effect and ride-through

The grid interactive inverters must follow the voltage and frequency characteristics presented on the distribution line. When the utility supply fails or the inverter is cut off from the grid, the grid-tied PV inverters will lose the reference voltage. This is defined as the islanding effect.

This islanding effect is similar to another transient phenomenon called ride-through. In Figure 3.9, after the grid voltage is converted from AC to DC, the DC voltage will be transformed into AC by the inverter to supply the drive machine. The AC grid might suffer blackouts for 1–2 seconds in some special cases. Without any AC grid detection, the energy stored in the DC capacitors will be drained very quickly. For example, if the DC voltage is 6000 V, DC capacitance is 3000 μF, and the motor is 1250 kW, without grid support the DC-bus energy will be exhausted within

$$t = \frac{\frac{1}{2}CU^2}{P} = 43.2 \,(\text{ms}) \tag{5.11}$$

Then the DC-bus voltage will reach zero and the motor will stop. Without any further actions, in-rush current will emerge in the front-end rectifier when the AC grid recovers. If the grid only suffers blackouts for several seconds, the motor drive system can sustain the DC-bus voltage by decelerating, thereby feeding the kinetic energy back to the DC bus, which is regarded as an induction generator instead of a motor. This transient process is known as ride-through [10]. However, if the grid is devastated, feeding the kinetic energy of the motor back to the DC bus is not sufficient to hold the DC-bus voltage. In this case, the motor needs be halted and all the semiconductor switches and relays need to be shut down. The whole system can only be allowed to restart with the proper start sequence when the AC grid recovers.

Grid blackouts make the grid-tied system lose the voltage reference, whether they last several seconds or longer. For the grid-tied PV system, when the grid is cut off, the control system must observe the islanding situation. In [11], the reference sinusoidal waveforms are overlapped with very small asymmetric signals. When the system is connected to the grid, the distorted current caused by the asymmetric signals is detectable. When islanding happens, the distorted current will disappear. After detection of the islanding effect, the most effective way presently to cope with the islanding effect is to halt the PV system.

Compared to the string inverter, the micro-inverter could handle the islanding effect more effectively. Implementation of micro-inverters disperses the risk of the islanding effect. First, there is only a small possibility that the islanding affect will happen to all inverters, compared to the single-string inverter. Second, a low-power inverter can always be designed with relatively higher electric allowance than a high-power inverter, which will make the micro-inverter more resilient to the grid impact.

5.2.2.4   In-grid control

When the system is connected to the grid, the electrical impact will happen due to the difference of the voltage amplitude and phase. This happens not only to the solar grid-tied system, but also to the wind energy system, a major rival of solar energy in alternative energy systems.

5.2.3   Wind energy systems

In the wind energy system, an induction motor or a permanent magnet motor are the main components. There are two major types of induction machines. One is the squirrel cage machine, whose rotor windings are totally short-circuited, and the other is the wound-rotor induction machine, whose rotor windings can be connected to an external circuit through sleeve rings. The wound-rotor induction machine is very suitable for wind power application where the rotor needs to be controlled in order to stabilize the generated power. When the rotor and the shaft work together to feed the energy back to the stator, the machine is named a doubly fed generator, which is widely used in wind energy systems [12].

An equivalent circuit of the induction motor is described below, where no external power supply is connected to the rotor. Here L_s and R_s are the stator leakage inductance and resistance, respectively, L'_r and R'_r are the equivalent leakage inductance and resistance of the rotor, respectively, L_m is the excitation inductance, R_m represents the iron loss of the motor, while R_s and R'_r stand for the copper loss; s is the slip. Figure 5.9 shows the T circuit and further simplified circuit of the induction motor.

The power consumed by $R'_r(1 - s)/s$ represents the shaft power. Neglecting the iron loss, we can obtain the shaft power as

$$P = \frac{U_s^2}{\omega^2(L_s + L'_r)^2 + \left(R_s + \dfrac{R'_r}{s}\right)^2} \times \frac{R'_r(1 - s)}{s} \tag{5.12}$$

Therefore the electromagnetic torque is

$$T = \frac{P}{\Omega} = \frac{U_s^2}{\omega^2(L_s + L'_r)^2 + \left(R_s + \dfrac{R'_r}{s}\right)^2} \times \frac{R'_r(1 - s)}{s\Omega} \tag{5.13}$$

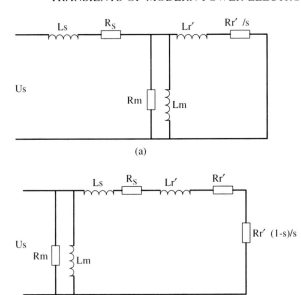

Figure 5.9   Equivalent circuit of the induction motor: (a) T circuit and (b) simplified circuit.

Here Ω is the mechanical angular speed of the motor. Assuming that the load torque stays the same, and the inductive impedance $\omega(L_s + L_r')$ is much higher than the resistance, Equation 5.13 can be further simplified as

$$T = \frac{P}{\Omega} = \frac{U_s^2}{\omega^2(L_s + L_r')^2} \times \frac{R_r'(1/s - 1)}{\Omega} \tag{5.14}$$

When the rotor is externally connected to a resistor, the slip will increase, which changes the motor speed. However, the loss on the external resistor is also considerable. With the power electronic converter, the rotor windings can be connected to external power supplies instead of a resistor, which totally changes the equivalent circuit of Figure 5.9 to Figure 5.10.

The rotor is supplied with some specific current at the rotor frequency (slip frequency). The magnitude and phase of U_s and I_r' will determine the magnetizing current through L_m and thus the relevant back electromotive force (EMF). When the back EMF is higher than U_s, or is leading U_s, the motor will act as a generator.

A typical wind energy system is shown in Figure 5.11.

If the loss is not included, the power generated by the stator is the summation of the wind power and the rotor power, that is,

$$P_s = P_{mech} + P_r \tag{5.15}$$

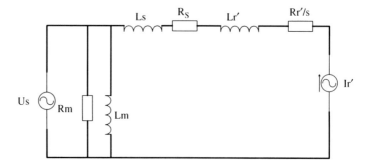

Figure 5.10   Doubly fed induction generator.

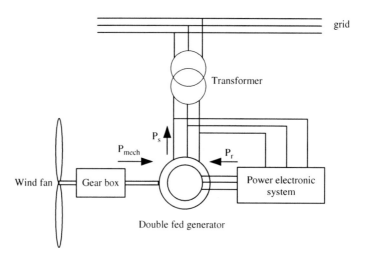

Figure 5.11   Wind energy system.

Based on machine theory,

$$P_r = sP_s$$

$$P_{mech} = (1 - s)P_s \qquad (5.16)$$

If $s > 0$, the rotor will absorb power from the stator, therefore the mechanical power delivered to the grid is $(1 - s)P_s$. If $s < 0$, the rotor speed is higher than the nominal speed, therefore the mechanical power to the grid is $(1 + |s|)P_s$. In this case the energy flows from the rotor to the grid. Therefore the power electronic converter should be bidirectional.

Compared to the squirrel induction generator, the doubly fed induction generator has many advantages. The most important one is that the power electronic converter only handles a small portion of the output power, which can be seen in

Equation 5.16. Therefore the cost and size of the drive system will be significantly less. However, the demerits are also obvious. In this system, two fatal processes need to be addressed:

1. **In-grid process:** this process will transfer the current impact to the motor and the grid. With modulation of the rotor current and frequency, the voltage of the stator could match the grid voltage amplitude and phase, therefore the process of connecting the stator to the grid is expected to be smooth. However, this requires highly accurate detection. Any small difference between the motor voltage and grid voltage will cause a large in-rush current.

2. **Triple-state transfer:** the wind speed changes all the time. Therefore the rotor speed could be less than, equal to, or larger than the nominal speed of the doubly fed generator. As shown in Equation 5.16, when the rotor speed is jittering around the nominal speed, the direction of the energy through the power electronic system changes frequently and adds complexity to the control.

The above conflicts can be mitigated if the topology shown in Figure 5.12 is adopted. In [13], the free wind speed is tested, under which the maximum power that the motor can generate is calculated. For VSI 1, the MPPT control should be adopted and for VSI 2 the constant voltage control is recommended to maintain the DC bus voltage. A DC-bus capacitor placed between the motor and the grid will buffer the in-rush energy from either side.

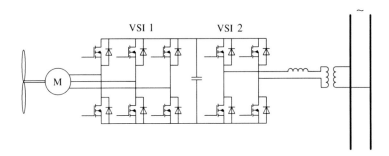

Figure 5.12   Wind energy system using an induction motor.

5.3   Power electronics, alternative energy, and future micro-grid systems

When alternative energy and the electric grid are tightly coupled, micro-grid topology comes into play. A micro-grid is a localized group of electricity sources and loads that operates as a stand-alone electrical system. This system can also be connected to and synchronized with the traditional centralized grid through a single point of common coupling with the macro-grid as needed, but can be disconnected and function autonomously as physical and/or economic conditions dictate [14]. At present, economic, technological, and environmental incentives force those centralized generating facilities to give way to smaller, more distributed generation. This is particularly important when grid blackouts happen, although micro-grid generation across the world has not yet reached significant levels. It is expected that the situation will change rapidly when distributed power generation and transmission are increasingly addressed.

When renewable energy and battery chargers for PHEVs are tied together, an environmentally friendly micro-grid system emerges to realize a green charging station, as shown in Figure 5.13.

At peak time, the local battery and PV panel will work together to charge the battery pack in the vehicle. At off-peak time, the PV panel will charge the local battery as a backup for peak time. In Figure 5.13a, one central processing unit (CPU1) will be responsible for the in-grid charging station and the other, CPU2, will operate the off-board part of the charger since bidirectional energy flow is required. Their communication can be wireless. In addition, CPU1 will communicate with the local PC to display the energy usage. In emergencies, different charging stations will be connected to the community central station, as shown in Figure 5.13b. Individual stations will report their energy excess or insufficiency and broadcast their willingness to trade in any energy from their neighborhoods. The community center will help to electrically connect those individuals who would like to exchange energy internally. Such a topology is also suitable for public charging stations, whose difference lies in the higher charging power.

The effectiveness of this micro-grid infrastructure depends on the integration of bidirectional inductive or wireless chargers, solar energy panels, and the electric grid. Operation of the system during dynamic processes is also important. For instance, the collapse of the solar energy system and/or electric grid from potential risks should not be a barrier to the use of charging stations. In addition, a grid-tied solar energy charging system is a promising strategy for complementing the grid and achieving the following. First, when massive renewable energy systems are working simultaneously, the quality of grid electricity is a concern. Second, a user friendly charging system should allow a householder to know the basic statistics of energy storage, delivery, and demand. The voltage, current, SOC, and temperature of the battery are among the most important parameters. Third, both the information and power flow should be bidirectional. This is particularly important when some householders need energy from others at the time of blackouts.

In the home charging station, wireless communication can be used to exchange information between the two sides of the inductive charger. The potential communication protocols must establish a wireless connection between the vehicle on-board unit (OBU) and the off-board facilities at home. Two communication protocols are in use in similar applications for charging the vehicles. The ZigBee wireless protocol is currently in use to connect the OBUs

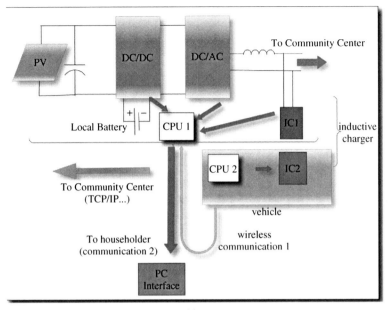

(a)

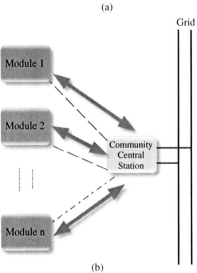

(b)

Figure 5.13   Future green charging station for micro-grid community: (a) one charging system in the house and (b) connections of charging systems in the community.

of some vehicles to a charging station [15]. ZigBee is also utilized to wirelessly connect in-house appliances, thermometers, and other equipment. The other protocol is dedicated short-range communication (DSRC), which is used to connect the OBUs of many vehicles to the roadside units (RSUs) [16].

The conceptual diagram of Figure 5.13 could be realized as Figure 5.14. On the house side, the PV panel, battery 2, the primary side of the charger, and the grid are all connected to the DC bus, C_1, which makes it a typical multi-source system. Full-bridge topology is used for the charger to realize bidirectional energy flow. Different from Figure 5.13a, the input of the charger is DC instead of AC, which eliminates the PFC circuit required in AC-supplied battery chargers. A controlled AC–DC converter is used to feed energy back to the AC grid and automatically realize the unity power factor. Since most energy coming from the solar panel goes to the vehicle battery directly, the power rating of the system is determined by the PV power, no longer restricted by the AC side. Therefore the power rating can be much higher than the present commercial 110 V AC charger (Level 1, around 1 kW [17]) without imposing any extra stress on the grid.

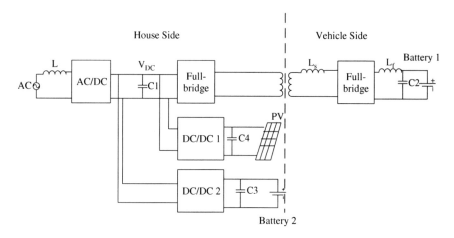

Figure 5.14   Future green charging station for micro-grid community.

The first problem encountered is reliability. When the islanding effect happens, the DC-bus voltage on C_1 can continuously increase when battery energy is fed back to the grid (Mode 1), and decrease when the grid is involved in charging the battery (Mode 2), or even stand still when only solar energy is used to charge the vehicle battery (Mode 3). For Modes 1 and 2, the abnormal DC-bus voltage will directly reveal the islanding effect. The direct protection is to shut down all the devices and halt all charging/discharging activities. In Mode 3, since the AC–DC converter does not handle any power flow, this converter could be shut down directly. Therefore the topology in Figure 5.14 is highly immune from the islanding effect, which is the major advantage.

Equipping the solar energy-based micro-grid system will lead to considerable economic benefits. Table 5.1 lists the solar radiation of Flint, MI, in the United

Table 5.1   Solar distribution in Flint, MI, USA.

Month	Solar radiation (kWh/m^2/day)	Energy (kWh/day)	E_{extra} (kWh/month)	Benefits ($)
1	2.1	14.4	105	8.4
2	3.3	22.7	327	28.1
3	4.5	30.9	617	49.3
4	5.9	40.6	897	71.8
5	7.1	48.8	1172	93.7
6	7.8	53.6	1278	102.2
7	7.8	53.6	1321	105.6
8	6.9	47.4	1128	90.3
9	5.3	36.4	762	61
10	3.6	24.8	428	34.2
11	2.1	14.4	102	8.16
12	1.6	11	0	0
Total	–	–	8137	653

States. The minimum solar radiation is 1.6 kWh/m^2/day. Usually the efficiency of the PV panel is less than 20%. Assuming all the energy comes from solar panels, the PHEV battery is rated at 11 kWh, daytime lasts 8 hours, and the charger efficiency is 80% (considering it is an inductive charger and including the battery loss), then the whole area of the solar panel is $11/1.6/0.8/0.2 = 43$ m^2. In this case, the charging time is 8 hours. If the vehicle battery charging time is set as 2 hours, then the power rating of the charger needs to be around 6 kW, and the required area for the solar panel is 172 m^2.

In any month of the year other than December, the house always has excessive energy to store or feed back to the grid or supply other houses. According to Table 5.1, each house has extra electricity of 8137 kWh per year, which is equivalent to $653. If one community has 200 houses, the total economic benefits will be $130 600. Considering the 11 kWh daily electricity to charge the PHEV, the total energy saving will be $194 432.

5.4   Dynamic process in the multi-source system

The infrastructure needed to accommodate multiple renewable or distributed energy sources is surely more complicated than traditional centralized energy generation and distribution. Compared to the centralized power system, the distributed power system uses different kinds of power generators, especially the flexible renewable power sources. Also, because these power sources are distributed locally, the cost of power transmission is reduced, thus allowing applications in many places where a centralized power system is not feasible, like an aircraft, aerospace system, or remote locations. Figure 5.14 shows a typical distributed power system where solar/wind energy, grid, and battery coexist in a single system. How to coordinate those power sources to attain a fast dynamic response is the responsibility of power electronics.

Different from AC distributed power systems, DC distributed power systems use a DC bus to connect each source and load, as shown in Figure 5.14. Compared to the AC system, the DC distributed system have many advantages: for example, no need to control frequency stability or reactive power, lower losses with less skin effect, and so on. In addition, consumer electronics, LED lighting, and variable speed motors are more conveniently powered by a DC bus. Also, a DC bus provides more benefits for some novel power sources, for example, batteries, fuel cells, and PV panels.

For a single-source-supplied DC bus, the voltage-controlled converter can easily manage its voltage. However, when the DC bus is controlled by several different sources, if each source runs in the voltage-controlled mode, the power distribution among the different sources will be totally uncontrollable. Some of the sources might be overburdened while others underutilized.

To avoid this problem, a feasible way is to let a more robust source maintain the DC-bus voltage while other sources fill the power gap, or more specifically, the current gap. Thus, one converter adopts voltage regulation while the others use current regulators to dynamically reach the targeted current.

For example, in Figure 5.15a, a multi-source system including a battery and a fuel cell is displayed. A fuel cell typically produces electricity through electrochemical reactions using hydrogen and oxygen, and produces water as a by-product. However, pure hydrogen does not exist in nature. One common approach is to break down water through a process of electrolysis to get pure hydrogen. In this process the consumed energy comes from external electricity provided by other sources, like solar energy, wind energy, coal, nuclear energy, or liquid fuel.

Fuel cell vehicles have attracted much interest in recent years because they offer zero emissions. In order to compete with the engine-propelled vehicle, the fuel cell must provide similar or even better dynamic response, drivability, safety, performance envelope, and so on. Usually the energy generated by the fuel cell is not stable (total energy is dependent on the rate of current drawn from the fuel cell). A subsequent power electronic converter is required to stabilize the output voltage or extract maximum power from the fuel cell. Meanwhile the response time of the fuel cell is slower, hence the battery and/or ultracapacitor are needed to provide the power gap in the dynamic process. For example, if the rated power of the fuel cell is 20 kW while in the acceleration process a total of 29 kW is required, the 9 kW gap should be provided by the battery, as shown in Figure 5.15b, or a combination of battery and ultracapacitor, as shown in Figure 5.15e. Therefore DC/DC 1 is working at the constant voltage mode and DC/DC 2 and DC/DC 3 are working at the power (current) mode.

The energy generated from the fuel cell and battery together will drive the motor, as shown in Figure 5.15b. The motor will drive the wheels through the differential. DC/DC 1 is unidirectional because the fuel cell does not receive electrical energy from external sources. The braking energy of the motor can be

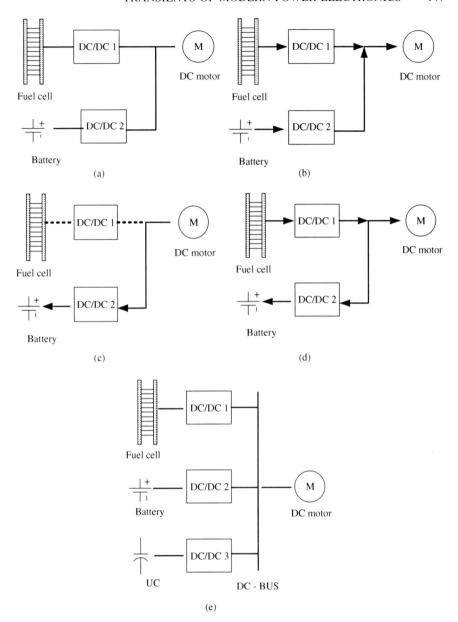

*Figure 5.15   Fuel cell hybrid vehicle: (a) schematic of a fuel cell hybrid vehi-
cle, (b) in normal operation modes, (c) in braking process, (d) in low SOC, and
(e) multi-source system.*

fed back to the battery and ultracapacitor as shown in Figure 5.15c,e. When the SOC of the battery is low, the fuel cell can provide the energy to charge the battery, therefore DC/DC 2 and 3 must be bidirectional.

One feasible way to allocate different power to the coexisting sources in Figure 5.15 is by the wavelet transform. Wavelet transform-based control has been used to provide the required power from a system including a battery, fuel cell, and ultracapacitor [18]. When the on-line power demand changes, the power distribution among the fuel cell, battery, and ultracapacitor also changes, as shown in Figure 5.16. The advantage of the wavelet transform is its efficacy in detecting high-frequency energy components. In this approach, the high-frequency components (including positive and negative power) are supplied

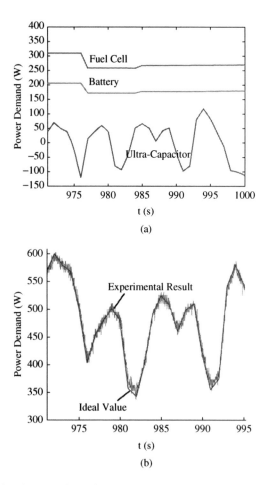

Figure 5.16   Wavelet transform-based multi-source system: (a) simulation results and (b) comparison of the simulation and experiment. © [2010] ELSEVIER. Reprinted, with permission, from Journal of Power Sources.

(or absorbed) by the ultracapacitor, the lower frequency components are supplied or absorbed by the battery, while the average power is supplied by the fuel cell. Through this control mechanism, the life of the battery can be significantly extended while the system dynamic performance is greatly enhanced.

5.5 Speciality of control and analyzing methods in alternative energy systems

Besides solar energy, wind energy and fuel cells, wave/tidal energy, nuclear energy, and biochemical energy are other typical alternative energy resources. Although renewable energy sources are abundant, the overall utilization rate of these sources is still very low (in the United States the ratio of all renewable energy sources to total electricity generation is less than 2%). In their applications, the following aspects need to be addressed:

1. **Instability:** for solar and wind energy, despite their abundant nature, their instability needs to be alleviated. Solar energy is greatly affected by the solar intensity, which is usually strongest at noon and weakest at night. Season and weather are also major influential factors that will significantly alter the output power capability of the solar energy system. Therefore a power electronic converter with MPPT control is needed.

   Instability is also a major issue in the wind energy system. The speed of the wind is forever changing, which alters the operation mode of the wind turbine blades. The altered blade speed leads to a continuously changing motor speed. Without power electronic converters, the quality of the generated power cannot meet the demands of customers.

2. **Decreased reliability:** the introduction of power electronic converters puts the application of alternative energy into practice. Up to now power electronics have been the best way to coordinate the multiple energy sources in one system. However, the introduction of power electronics will decrease the reliability of the system. This is particularly true in the high-voltage and high-power applications of renewable energy systems where many potential failures could happen due to the large stress and increased number of components. In this case, microconverters can be a good alternative, which mitigate system-level failure through the utilization of many parallel-connected converters.

3. **In-grid control:** at present, the role of fossil energy is still dominant in electricity generation and other applications. Most alternative energy systems today are merely to provide some extra energy to the grid and act as a backup for the fossil-based energy system. Therefore their in-grid control is greatly appreciated. The impact on the grid should be minimized in both the transient and steady states. A power electronic converter needs to be installed between the grid and stand-alone energy generating system.

Based on the real-time sampling and detection of grid status, the characteristics of the generated electrical parameters of the converter should be "grid friendly."

The cost and reliability of the alternative energy are the major constraints at the present time. Presently, alternative energy still cannot compete with gasoline or coal on price. Also, "alternative" does not always mean "renewable," for example, the fuel cell. The environmental impact of different alternative energy sources is still arguable, such as nuclear power. However, the application of alternative energy makes the future micro-grid system possible. The solar energy-based house charging station for the future micro-grid community driving PHEVs is, as explained in this chapter, a very typical example. The emergence of the micro-grid not only releases customers from potential grid blackouts, but also shrinks the dependence on fossil energy and eventually becomes environmentally friendly.

5.6   Application of power electronics in advanced electric power systems

Power electronic converters are now increasingly used in modern power systems. Due to the nature of high power, high voltage, and high current, the switching frequency in these applications is typically low, in the hundreds of hertz range. Hence, transients in these types of applications are critical for system stability, safety, and reliability.

A typical example of the application of power electronics in modern power systems is the flexible alternating current transmission system (FACTS), which is defined by the IEEE as "a power electronics-based system and other static equipment that provide control of one or more AC transmission system parameters to enhance controllability and increase power transfer capability" [19]. The FACTS includes series compensation and shunt compensation. In series compensation, the FACTS is connected in series with the power system working as a controllable voltage source. Since series inductance occurs in long transmission lines, which causes a large voltage drop when a large current flows, series capacitors are connected to compensate this voltage drop. In shunt compensation, the power system is connected in parallel with the FACTS. It works as a controllable current source.

The prevalent FACTS applications are: thyristor-controlled series compensator (TCSC), static synchronous compensator (STATCOM), static synchronous series compensator (SSSC), convertible static compensator (CSC), unified power flow controller (UPFC), thyristor-switched capacitor (TSC), thyristor-controlled reactor (TCR), thyristor-switched filter (TSF), thyristor-controlled phase regulator (TCPR), dynamic voltage restorer (DVR), subsynchronous resonance dumper (SRD), superconducting magnetic energy storage (SMES) system, battery energy storage system (BESS), and solid state circuit breaker (SSSB). The traditional TCSC, TSC, and TCR focus on compensating the long-term steady state reactive power. With the development of modern power electronic devices, the STACOM, UPFC, SMES, etc., have gradually come to dominate in FACTS application due to their fast response and small impact on the grid. We will briefly discuss a few applications in this section.

5.6.1   SVC and STATCOM

A static Var compensator (SVC) is a widely used electrical device in the electric grid for providing prompt reactive power, which is critical in high-voltage electricity transmission networks. As part of the FACTS device family, SVC plays the role of regulating voltage and stabilizing the system, by compensating reactive power in the grid. The SVC scheme is shown in Figure 5.17.

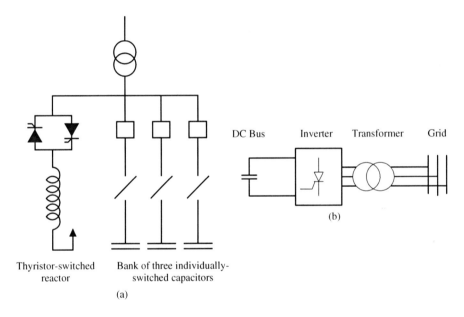

Figure 5.17   Principle of SVC and STATCOM. (a) SVC (b) STATCOM.

Normally a SVC has two sets of equipment, one of which is an inductor and other a capacitor. If the reactive power on the grid is detected as capacitive, the SVC will switch the inductor to compensate the reactive power, otherwise it will use the capacitors. Therefore it could be regarded as a special example of the PFC, which was discussed in Chapter 4.

With the acceptability of GTOs and IGBTs among the utilities, STATCOMs are also becoming popular. A STATCOM is defined as "a regulating device used on AC electricity transmission networks. It is based on a power electronic voltage source converter and can act as either a source or sink of reactive AC power to an electricity network. If connected to a source of power, it can also provide active AC power" [20]. The STATCOM is a typical member of the FACTS. The simplest STATCOM structure is a three-phase, two-level DC–AC inverter. Its theory has been described in Equations 5.5–5.9.

Although SVC and STATCOM both belong to the FACTS, their compensating ability is different. As shown in Figure 5.17, the SVC comprises two parts, namely, a TSC and TCR. They are connected in parallel and together compensate the system. Their current and voltage are highly coupled, as shown in Figure 5.18a for the voltage–current characteristics.

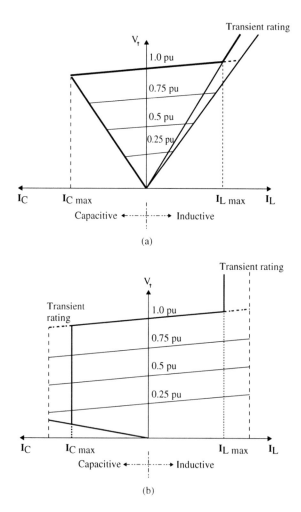

Figure 5.18   V–I characteristics of the SVC and STATCOM: (a) for SVC and (b) for STATCOM. © [2005] IEEE. Reprinted, with permission, from IEEE Transactions on Power Delivery.

Through electronic processing of the voltage and current waveforms in a voltage source converter, STATCOM will either provide or absorb the targeted active and/or reactive power on the power lines. Figure 5.18b shows the $V - I$ characteristics of a STATCOM, which controls its output current over the rated maximum capacitive or inductive range independently, regardless of the amount of grid voltage. Compared to the SVC, the STACOM is more flexible.

Both SVC and STATCOM will be impacted when they are switched to the grid. The reason is the existence of the capacitors. Therefore, before starting up, the capacitors need to be charged accordingly. Only when the DC-bus capacitor is charged to the rated voltage is the STATCOM allowed to start.

5.6.2 SMES

A SMES system is another typical application of power electronics. In addition to the power electronic converter, it has a superconducting coil and cryogenically

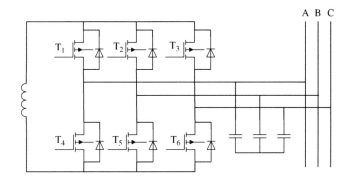

Figure 5.19   CSI scheme.

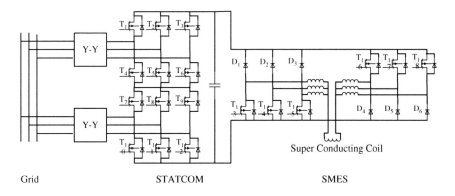

Figure 5.20   STACOM + SMES. © [2003] IEEE. Reprinted, with permission, from IEEE Industry Applications Magazine.

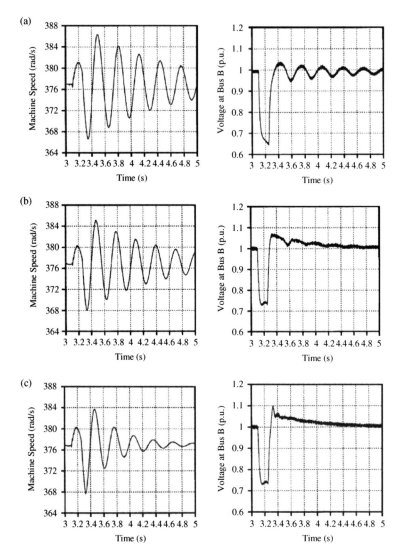

Figure 5.21 Dynamic response to AC system oscillations: (a) no STAT-COM/SMES, (b) STATCOM only, and (c) STATCOM/SMES. © [2003] IEEE. Reprinted, with permission, from IEEE Industry Applications Magazine.

cooled refrigerator. The very low temperature of the superconducting coil guarantees that the coil current will not decay and the magnetic energy can be stored forever once it is charged. The topology is shown in Figure 5.19 where a current source inverter (CSI) is used. Through discharging the coil, energy can be fed back to the grid. Power electronics play the key role of transforming the direct current into alternating current in the SMES.

SMES utility applications have many advantages when used as the FACTS, for example, prompt response in milliseconds, high power ratings at multi-megawatts, high efficiency (the loss in the coil is nearly zero), and flexible bidirectional control. Therefore SMES systems can improve system reliability, enhance dynamic stability, and implement area protection. However, their cost is still high. At present, SMES is used as an energy storage device for a short duration.

In [21], STATCOM and SMES are tied together. A STATCOM can only handle reactive power. Absorbing/generating active power through a STATCOM will tend to vary the DC-bus voltage thereby endangering the inverter. In this case, adding SEMS will allow the STATCOM to generate or absorb active and reactive power simultaneously. The schematics are shown in Figure 5.20.

In the simulation in [21], a three-phase fault created at $t = 3.1$ seconds was cleared at $t = 3.25$ seconds. The load was a three-phase machine. When there is no STATCOM/SMES connected to the grid, the fault will let the machine speed oscillate by itself (Figure 5.21a). When a STATCOM only is connected, the response is as given in Figure 5.21b. Since the STATCOM is used for compensating the reactive power, it is not effective in damping oscillations. Therefore at $t = 5$ seconds, the machine speed is still oscillating. In Figure 5.21c, SMES is added to absorb the active power and the STATCOM is used to handle the reactive power. At $t = 5$ seconds, the machine is already in the steady state.

References

1. Huang, S.J. and Pai, F.S. (2001) Design and operation of grid-connected photovoltaic system with power-factor control and active islanding detection. *IEE Proceedings on Generation, Transmission and Distribution*, **148** (3), 243–250.
2. Goetzberger, A. and Hoffmann, V.U. (2005) *Photovoltaic Solar Energy Generation*, Springer, Berlin.
3. Peiyu, W., Boxue, T., Housheng, Z., and Yanlei, Z. (2010) Research on maximum power point tracker based on solar cells simulator. 2nd International Conference on Advanced Computer Control, Vol. 1, pp. 319–323.
4. Femia, N., Petrone, G., Spagnuolo, G., and Vitelli, M. (2005) Optimization of perturb and observe maximum power point tracking method. *IEEE Transactions on Power Electronics*, **20** (4), 963–973.
5. Libo, W., Zhengming, Z., and Jianzheng, L. (2007) A single-stage three-phase grid-connected photovoltaic system with modified MPPT method and reactive power compensation. *IEEE Transactions on Energy Conversion*, **22** (4), 881–886.

6. Taherbaneh, M. and Menhaj, M.B. (2007) A fuzzy-based maximum power point tracker for body mounted solar panels in LEO satellites. IEEE/IAS Industrial and Commercial Power Systems Technical Conference, pp. 1–6.

7. Yi-Bo, W., Chun-Sheng, W., Hua, L., and Hong-Hua, X. (2008) Steady-state model and power flow analysis of grid-connected photovoltaic power system. IEEE International Conference on Industrial Technology, pp. 1–16.

8. Backlund, B. and Carroll, E. (2006) Voltage Ratings of High Power Semiconductors, Product information from ABB Switzerland Ltd, p. 10.

9. Kutkut, N. and Hu, H. (2010) Photovoltaic microinverters: topologies, control aspects, reliability issues, and applicable standards. IEEE Energy Conversion Conference and Exposition, University of Central Florida (tutorial materials).

10. Ng, C.H., Ran, L., and Bumby, J. (2008) Unbalanced-grid-fault ride-through control for a wind turbine inverter. *IEEE Transactions on Industry Applications*, **44** (3), 845–856.

11. Kim, J.-H., Kim, J.-K., Jung, Y.-C. *et al.* (2010) A novel islanding detection method using Goertzel algorithm in grid-connected system. International Power Electronics Conference, pp. 1994–1999.

12. Seman, S., Niiranen, J., Kanerva, S. *et al.* (2006) Performance study of a doubly fed wind-power induction generator under network disturbances. *IEEE Transactions on Energy Conversion*, **21** (4), 883–890.

13. Abo-Khalil, A.G. and Lee, D.-C. (2008) MPPT control of wind generation systems based on estimated wind speed using SVR. *IEEE Transactions on Industrial Electronics*, **55** (3), 1489–1490.

14. Wikipedia (2011) http://en.wikipedia.org/wiki/Microgrid (accessed March 25, 2011).

15. Liang, L., Huang, L., Jiang, X., and Yao, Y. (2008) Design and implementation of wireless smart-home sensor network based on ZigBee protocol. International Conference on Communications, Circuits and Systems, pp. 434–438.

16. Mak, T.K., Laberteaux, K.P., Sengupta, R., and Ergen, M. (2009) Multichannel medium access control for dedicated short-range communications. *IEEE Transactions on Vehicular Technology*, **58** (1), 349–366.

17. Scholer, R.A., Maitra, A., Ornelas, E. *et al.* (2010) Communication between Plug-in Vehicles and the Utility Grid, SAE Paper Number 2010-01-0837, doi: 10.4271/2010-01-0837. http://papers.sae.org/2010-01-0837/.

18. Zhang, X., Mi, C., Masrur, A., and Daniszewski, D. (2008) Wavelet based power management of hybrid electric vehicles with multiple onboard power sources. *Journal of Power Sources*, **185** (2), 1533–1543.

19. Edris, A., Adapa, R., Baker, M.H. *et al.* (1997) Proposed terms and definitions for flexible AC transmission system (FACTS). *IEEE Transactions on Power Delivery*, **12** (4), 1848–1853.

20. Wikipedia (2010) http://en.wikipedia.org/wiki/STATCOM (accessed March 25, 2011).

21. Arsoy, A.B., Liu, Y., Ribeiro, P.F., and Wanf, F. (2003) Static-synchronous compensators and superconducting magnetic energy storage systems in controlling power system dynamics. *IEEE Industry Applications Magazine*, pp. 21–28.

6

Power electronics in battery management systems

6.1 Application of power electronics in rechargeable batteries

Rechargeable batteries are nowadays widely used in consumer electronics, uninterruptable power supplies, electric and hybrid electric vehicles, and renewable energy systems. The size of the battery used in different applications can vary significantly. For example, PHEVs require larger battery packs and electric drivetrain components compared to regular HEVs. EVs require even higher battery ratings since all the driving range is provided by the onboard battery. The battery size for HEVs is around 1 to 2.5 kWh, while the battery of PHEVs and EVs can be anywhere from 7 to 40 kWh [1]. On the other hand, batteries in consumer electronics can be just a few watt-hours or a few tens of watt-hours.

There are many types of rechargeable batteries available for various applications today. The most popular rechargeable batteries include the lead–acid battery, nickel metal hydride (NiMH) battery, and lithium-ion battery. In particular, lithium-ion batteries have been the center of focus currently for both consumer electronics and EV applications due to their high specific energy and high specific power [2–5].

Figure 6.1 shows the external characteristics of a lithium-ion iron phosphate battery. The figure shows that the battery State of Charge (SOC), is highly associated with its terminal voltage. It should also be pointed out that the two areas, one close to zero SOC and one close to 100% SOC, are of critical importance in managing the batteries. The terminal voltage of the lithium-ion battery should be strictly maintained within a safety range. In the example shown in Figure 6.1, the

Transients of Modern Power Electronics, First Edition. Hua Bai and Chris Mi.
© 2011 John Wiley & Sons, Ltd. Published 2011 by John Wiley & Sons, Ltd.

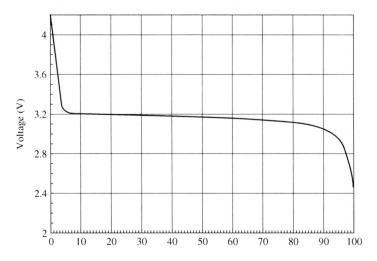

Figure 6.1   The battery voltage vs. DOD.

optimal battery voltage is between 2.8 and 3.8 V. Charging the battery above 3.8 V will likely cause overheating and potential damage to the battery. Discharging the battery below 2.8 V will also cause damage to the battery.

There are a few challenges in battery applications that are closely related to and can be addressed by power electronics, such as charge management, capacity retention, safety, protection, efficiency enhancement, and cell balancing. This chapter discusses three important aspects of the applications of power electronics in rechargeable batteries: charge management, cell balancing, and safety protection.

6.2   Battery charge management

The general charging method is designed based on the internationally recognized rule, the "Ampere-Hour-Rule," enacted in 1940: the total ampere-hours of charging should not exceed the expected ampere-hour rating of the battery. In reality, the charging rate is restrained by the temperature rise and generated gas in the battery. These phenomena are important in setting the minimum charging time.

The charging method plays a key role in maximizing battery performance. Proper battery-charging technique ensures battery safety and increases system reliability, thereby prolonging the lifetime of the battery. The primary requirement of the charging process is to provide a fast and efficient way without degrading the battery. All these functions are realized inside the charger.

6.2.1   Pulsed charging

Charging a battery with continuous current in the long term tends to polarize the battery, creates a vulcanized crystal in the lead–acid battery, and increases the ohmic effect in the NiMH battery [6]. Even for new types of batteries like the lithium-ion battery, continuing to pump energy into the cell faster than the speed of the chemical reaction can cause local overcharge conditions including polarization and overheating, as well as unwanted chemical reactions near the electrodes and damage to the cell. Therefore, continuous charging is supposed to shorten battery lifetime. Pulsed charging has been proposed to improve the battery performance. Specifically, in one charging period, this technology uses pulsed current to charge the battery, stops charging for a while, then charges again, and so on, as shown in Figure 6.2. The relaxation period between pulses give some leeway for the battery to let the ions diffuse and distribute evenly. This process normalizes the ion concentration in the battery and thereby prevents the negative effects experienced in constant current charging. Because of the equal charge distribution, battery performance as well as battery life are enhanced.

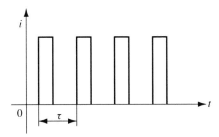

Figure 6.2   Pulsed charging.

Charging with a very high current in fast charging can lead to a high temperature inside the battery. Therefore, with pulsed charging, there will be enough time for the battery packs to cool down between pulses when the temperature exceeds the maximum allowed temperature.

6.2.2   Reflex fast charging

This technique consists of four different charging stages, graphically shown in Figure 6.3 [7]. The soft-start stage, which gradually increases current levels up to the user-selected fast charge rate, lasts for a few minutes. The soft-start stage is followed by a fast-charge stage, which is followed by a 2-hour C/10 topping charge. The topping charge is followed by a C/40 maintenance charge. (C is the rated ampere-hours of the battery.)

The reflex fast charging method is used in the NC2000, a device designed to intelligently charge either nickel–cadmium (NiCad) or NiMH batteries. This charging technique employs a four-stage charge sequence that provides a complete recharge without overcharging. In order to charge and discharge, the charger must be bidirectional.

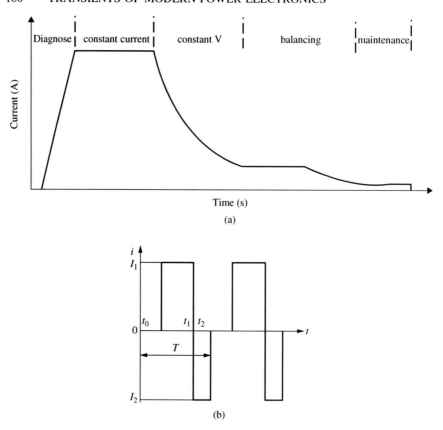

Figure 6.3   Graphical representation of average current levels during the four charge stages: (a) average mode and (b) one charging cycle.

6.2.3   Current variable intermittent charging

This method is illustrated in Figure 6.4 [8]. It sets an upper limit for the charging voltage in different intervals with different charging current. For example, in the first stage, a constant current is set to charge the battery. When the voltage approaches the upper voltage limit, the charge stage will be interrupted and the whole system enters an idle stage where the battery voltage drops gradually. This idle stage is followed by the next-cycle charging period with a smaller constant current setting. When the battery is close to the full state, the mode is changed to constant voltage charging.

The idle stage also leaves sufficient time for the battery to recombine its generated oxygen and hydrogen inside. At the same time, intermittent operation avoids the potentially dangerous temperature rise.

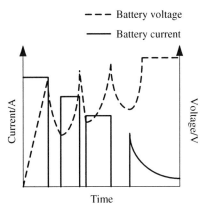

Figure 6.4   Current variable intermittent charging. © *[2000] IEEE. Reprinted, with permission, from ISIE 2000.*

6.2.4   Voltage variable intermittent charging

This method is shown in Figure 6.5, whose only difference from the above method is that the constant electrical parameter is not current, but voltage [8]. Charging current is left alone as long as it does not exceed the maximum settings that the system can handle. The charging current falls exponentially in every constant voltage charging stage.

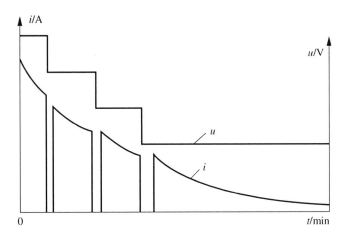

Figure 6.5   Voltage variable intermittent charging. © *[2000] IEEE. Reprinted, with permission, from ISIE 2000.*

6.2.5   Advanced intermittent charging

The advanced intermittent charging method combines the above charging method. It fixes the current amplitude and frequency and changes the current duty ratio to vary the average charging current [8]. The duty ratio in the fast charging stage is supposed to be larger than that in the voltage holding stage. This is illustrated in Figure 6.6.

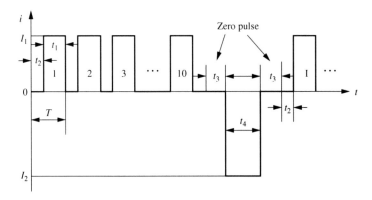

Figure 6.6   Voltage variable, current variable intermittent charging. © *[2000] IEEE. Reprinted, with permission, from ISIE 2000.*

6.2.6   Practical charging schemes

Practical charging schemes can be divided in two five different phases and described as follows:

- **Initial phase:** the initial charging phase is used to check the battery condition (such as voltage, initial SOC, health, etc.), and determine what charging method is the best fit for the battery. It is very common to include a trickle charging period in this phase, with a small amount of current imposed on the battery for a while before the operation mode changes to the fast charging phase [9].

- **Fast charging phase:** in this phase, the battery is charged at a relatively high rate based on the specification and condition of the battery. The methods discussed above can be applied in this phase, including constant current charging, constant voltage charging, and pulsed charging. The maximum charging current is determined by the limits specified by the manufacturer, including temperature, current, voltage and time, and the power capability of the charger. Some of the latest EV batteries can be charged at 2C or higher. Very fast charging (up to 15C, or 4 minutes to fully charge a battery pack) has also been reported, but the impact on battery life and long-term usable capacity is not yet well understood.

- **Absorption charging phase:** this phase takes a comparatively long time. The charging current is reduced in this phase. For example, C/140 should be applied for a time period about 1.6 times the initial charging phase to batteries like the Varta DRYMOBIL. Although this phase returns only a small amount of energy to the battery, it is essential for most batteries to prolong their life.

- **Float charging phase:** this phase maintains the battery in a fully charged state. The floating charging reverses all self-discharge processes and keeps the battery fully charged all the time when not in use. Some newer batteries, such as lithium-ion batteries, do not need this phase as they possess very little self-discharge.

- **Equalizing phase:** some types of batteries require a balancing phase to fully charge all cells that are connected in series. Usually this phase takes a very long time. The equalization techniques will be described in Section 6.3.

Appropriate charging can help prolong battery life and increase charging efficiency. For example, the battery is one of the most expensive components in a PHEV. The life expectancy of the battery is at least 10 years, compared to the 3–4 years typical of batteries used in consumer electronics. Therefore, prolonging battery life will make the HEV/PHEV more competitive.

Figure 6.7 illustrates the charge behavior of a lithium-ion phosphate battery module with a nominal voltage of 12.8 V. Figure 6.7a shows the depth of discharge (DOD) versus terminal voltage at different discharge rates. Figure 6.7b shows the charge current as a function of time for a typical pulsed charge profile. Figure 6.7c shows the terminal voltage during pulsed charge with charge current shown in Figure 6.7b. Figure 6.7d shows the battery SOC as a function of open circuit voltage at room temperature derived from the pulsed charging.

One study has shown that trickle charging followed by the fast charging stage is believed to be healthier for the battery than fast charging directly [10]. In [10], another healthy charging control, interrupted charge control (ICC), was proposed, whose operation modes are shown in Figure 6.8.

In Mode 1, the battery is charged with a constant current. At the end of Mode 1, the SOC of the battery is typically over 85%. Mode 2 is triggered when the battery voltage reaches the upper threshold. No current is employed to charge the battery; then the voltage falls due to the voltage drop across the internal resistance. Mode 3 is triggered when the open-circuit voltage drops below the lower threshold voltage. Here the battery is pulse charged with a series of pulse currents whose peak current is smaller than Mode 1. Mode 4 is triggered when the battery voltage again reaches the upper limit. The battery is expected to be fully charged at the start of Mode 4. The potential risk of being overcharged by the floating charge is greatly reduced.

Li *et al.* [11] comprehensively compared the impact of the different charging control algorithms on the battery cells, showing that pulsed charging imposes less stress on the battery cell than normal DC charging.

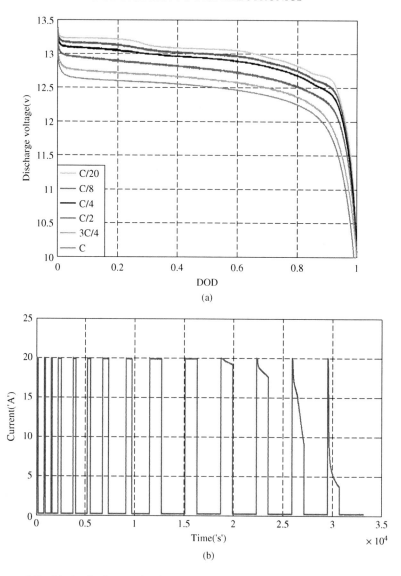

*Figure 6.7 Typical charge/discharge characteristics of lithium ion batteries.
(a) Voltage vs DOD during the discharge of lithium-ion battery at different discharge rates (the key refers to the curves in order, from top to bottom), at room
temperature. (b) Charge current during pulsed charge of lithium-ion battery. (c)
Voltage during pulsed charge using the charging current shown in (c). (d) Battery
voltage vs state of charge of a lithium-ion battery, calibrated through testing data.*

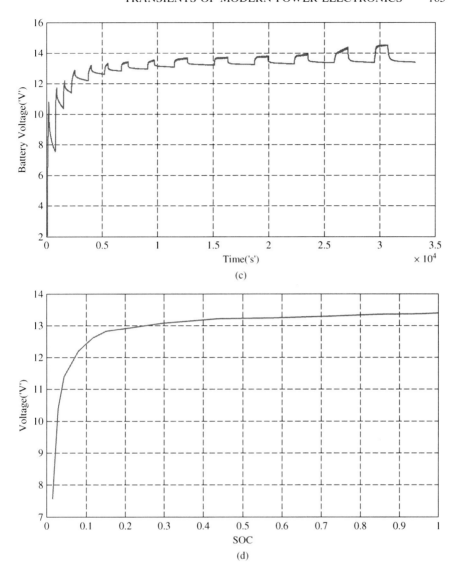

Figure 6.7   (continued)

Charging and discharging can both cause the temperature inside the battery to rise, which is determined by the heat balance between the amount of heat generated and that dissipated by the cell and associated cooling system. The temperature of a cell is a critical factor affecting battery life. Take the lead–acid battery as an example: it is believed that for every 10 °C rise in battery ambient temperature, the expected life is reduced by 50%. The temperature of the pack should be tightly monitored during charging, charge balancing, and discharging process.

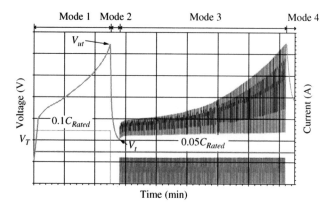

Figure 6.8 Interrupted charge control. © [2005] IEEE. Reprinted, with permission, from IEEE Transactions on Industrial Electronics.

When a cell is heated above a certain temperature (usually above 30–50 °C for the lithium-ion battery), the cooling system should be enabled. If the heat generated is more than what can be dissipated, the battery temperature will increase rapidly, which will further accelerate the chemical reactions. Eventually it will reach another thermal balance, which generates a pressure causing inevitable mechanical failures.

The above charging, balancing, and discharging process will need to be managed and regulated by power electronic circuits. For example, the ripple in the charger output should be maintained at a low level during the charging process.

6.3   Cell balancing

A typical PHEV system employs hundreds or thousands of battery cells in a battery pack. These cells are connected in series and parallel to achieve the required system voltage and current ratings. For example, a 320 V, 10 kWh battery string needs more than 100 cells of 3.2 V, 40 Ah batteries connected in series. The Tesla Motors Roadster EV adopted small-format cylindrical cells of a lithium-ion battery. There are 6800 cells in series and parallel connections. These are called 18 650 cells (i.e., 18 mm diameter, 65 mm height). The pack is rated at 375 V nominal voltage, which provides 53 kWh of energy and a maximum power of 200 kW, and weighs 450 kg [12].

From the point of view of power electronics and electric motors, a higher battery voltage is preferred because a smaller current is needed for the same power, therefore there is less stress and higher efficiency for cables, semiconductor devices, and motors.

When battery cells are connected in series to form a string, the available energy of the string is determined by the cell that has the least energy. Similarly, when charging the battery, the amount of energy that can be transferred to the string is determined by the cell that has the most energy left. Although battery manufacturers have been working on making battery cells consistent, there will still be small differences in the hundreds or thousands of cells that are used to form the string. On the other hand, the battery temperature can also be different inside a battery pack. Cells closer to the cooling inlet can experience a much lower temperature than cells that are close to the cooling outlet. The temperature difference for an air-cooled 10 kWh battery pack can be as high as 15 °C. It is well known that batteries exhibit very different internal impedance at different temperatures. Hence the cells at higher than normal temperature during hot weather will have more internal losses than other cells that are at normal temperature; and the cell at lower than normal temperature on cold days will also have more internal losses than other cells at normal temperature. This will result in these cells discharging quicker than the normal cells.

The example in Figure 6.9 shows a battery string with four cells at different SOC. On the left, all the cells have the same voltage, therefore they can be fully charged. On the right, the four cells do not exhibit the same voltage. If not managed, this string can cause a number of issues:

- During charge, if the individual cells are not managed, the cell that has the most energy (4.8 V as shown when the string voltage reached 14.6 V) could be overcharged and cause damage to the cell or even a thermal runaway.

- During discharge, the cell that has the least energy will be drained first, and at that point the string is no longer able to provide power even though some other cells still have a lot of energy. Further discharging of the string could potentially reverse the polarity of that weak cell and damage would occur.

Unmanaged battery strings start to show discrepancies over a period of time. Table 6.1 gives the voltage distribution inside an unmanaged battery string. $V_1 - V_4$ are the cell voltages of the battery string, and $T_1 - T_4$ are the temperatures, respectively. It can be seen that, due to the diversity of the battery cells, voltage imbalance occurs.

This issue can be managed by cell balancing circuits, sometimes referred to as equalizers. The methods of equalizing the batteries used in practice are as follows.

6.3.1 Applying an additional equalizing charge phase to the whole battery string

This method uses trickle current to charge the battery when the major charging process ends. The appropriate amount of current is imposed on the battery for a long time. For those batteries like the lead–acid, NiMH, or NiCd (unlike Li-ion) batteries, fully charged cells will not be damaged when further charged with these small amounts of energy. The fully charged cells transform this extra energy into heat, while the other cells in the string will be fully charged.

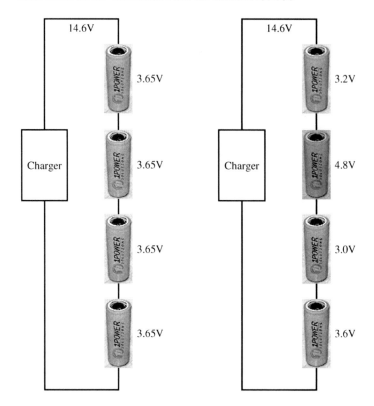

Figure 6.9   Battery imbalance in charging process.

Table 6.1   Voltage and temperature distribution inside the battery string.

ID	V_1	V_2	V_3	V_4	T_1	T_2	T_3	T_4	V
1	3.67	3.64	3.68	3.69	26	27	27	28	14.69
2	3.31	3.29	3.31	3.33	26	26	9	23	13.23
3	3.34	3.31	3.34	3.36	27	26	27	26	13.35
4	3.31	3.29	3.32	3.33	26	27	26	27	13.26
5	3.30	3.28	3.30	3.32	27	26	28	0	13.20
6	3.34	3.30	3.35	3.35	26	26	27	26	13.34
7	3.33	3.31	3.33	3.35	27	27	27	27	13.32
8	3.32	3.30	3.32	3.34	25	26	23	27	13.27
9	3.34	3.32	3.34	3.35	28	27	27	26	13.36
10	3.32	3.30	3.33	3.35	27	28	27	27	13.31
11	3.33	3.31	3.34	3.35	28	28	28	28	13.34
12	3.34	3.32	3.35	3.36	27	27	28	27	13.36
13	3.33	3.31	3.33	3.35	27	28	27	29	13.32
14	3.32	3.29	3.32	3.33	28	28	28	28	13.26
15	3.66	3.64	3.64	3.67	28	28	28	31	14.62

This method has several disadvantages:

1. Not suitable for lithium-ion batteries.

2. Equalization takes a very long time and will not reach 100% balance in most applications.

3. Difficult to adapt the method to the change of battery parameters over their lifetime (e.g., self-discharge rate increases over the lifetime).

4. High corrosion and aging rates because of heat generation in the cells during the equalizing phase.

5. No equalization during the initial charging process.

6. No equalization during the discharging process.

6.3.2   Method of current shunting – dissipative equalization

The principle of dissipative equalization is to pull energy from the fullest cells and dissipate it in a resistor or transistor [13]. Figure 6.10 shows the principle. A device such as a MOSFET is connected to each cell through resistors. When a cell is found to have a higher voltage or higher energy, the MOSFET is turned

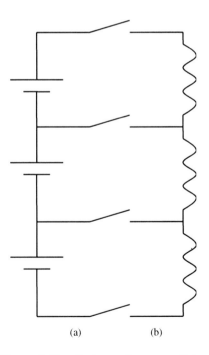

(a)   (b)

Figure 6.10   Resistor shunt equalizing.

on and the energy is consumed in the resistor. This is done continuously until all battery cells have the same voltage.

The advantages of this method are:

1. Simple and low-cost circuit.

2. Long wires are possible – no electromagnetic compatibility (EMC) problems, no high frequencies.

The disadvantages are:

1. Energy is dissipated into heat and therefore lost.

2. Higher equalizing currents require big and expensive heat sinks.

3. Continuous equalization is not recommended, because of high energy dissipation.

6.3.3 Method of switched reactors

The main disadvantages of dissipative equalizing are the loss of energy and the relatively low currents [13]. The method of switched reactors transfers energy from the cells with higher voltage to the lower charged cells instead of dissipating the energy. This method works bidirectionally, usually comparing two neighboring blocks. Less energy is "lost" and there is no need for big heat sinks. Figure 6.11 shows one circuit of a switched reactor in principle. The transistor next to the block or cell with higher charge is controlled with PWM. When switched on (phase 1) it draws current from this block through a reactor, which "stores this current." When switched off (phase 2), the neighboring block is charged with this small amount of stored energy through the diode.

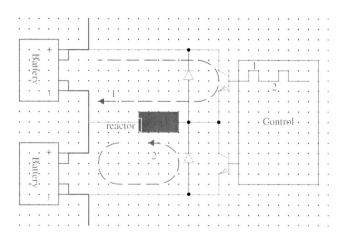

Figure 6.11   Switched-reactor equalizer.

The advantages of this method are:

1. High equalizing currents are possible.

2. Small loss of energy.

3. Bidirectional equalization: high voltage cells are discharged and low voltage cells are charged.

4. Continuous equalization is possible.

Its disadvantages lie in the higher complexity. Also, the energy can only be transferred between neighboring cells.

6.3.4 Method of flying capacitors

This method uses capacitors in combination with MOSFET switches [13]. Figure 6.12 shows the circuit principle. The switches just switch back and forth with a certain frequency. The capacitor between two blocks will reach the average voltage of these blocks. It discharges the cells with higher voltage in a first step and charges the cells with lower voltage in the second step.

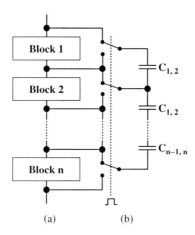

(a)   (b)

Figure 6.12   Capacitive equalizer.

The advantage is a very simple control algorithm, just switching with a certain frequency when equalization is enabled.

The disadvantages of the flying capacitors are:

1. Switching capacitors result in current peaks which cause unnecessary heat in the battery. Heat means faster aging.

2. Long equalization time when small capacitors are used.

6.3.5   Inductive (multi-winding transformer) balancing

Infeneon proposed a topology shown in Figure 6.13, which can be referred to as inductive balancing [13]. A multi-winding transformer is the key component to balance the cell voltage. In this method, the bidirectional equalization capability will assure that all battery cells have uniform charging and that the energy delivered by the battery string is maximized.

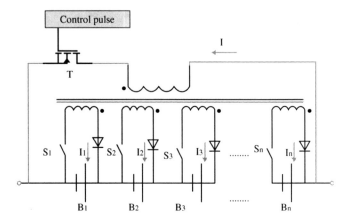

Figure 6.13   Multi-winding transformer-based balancing topology.

Initially, all switches are turned off. If a cell is detected to have a lower voltage, then the primary side of the transformer is first turned on. Current will build up in the primary winding. Then T is turned off and, in the meantime, the switch that is connected to the cell with the lowest voltage will turn on. The energy stored in the primary winding is transferred to this secondary winding and then to this cell. Once the energy is exhausted, the process is repeated until all cells have the same voltage.

6.3.6   ASIC-based charge balancing

1Power Solutions Inc. developed an approach based on application-specific integrated circuits (ASICs) as shown in Figure 6.14 [14]. The small IC is placed in parallel with each battery cell. The IC possesses properties similar to a Zener diode. When the voltage across the cell (hence the IC) is less than the preprogrammed voltage, the current through the IC is zero. Thus all charging current goes through the battery cell. When the voltage across any cell is more than the preprogrammed voltage, the IC starts to conduct current. Thus less current is entering the battery cell. The current through the IC will increase as the voltage further increases.

The advantage is that the IC is small and can be very accurate. The disadvantage is that the energy entering the IC is wasted as heat. A second disadvantage is that the IC needs a large heat sink if a large current needs to go through the IC.

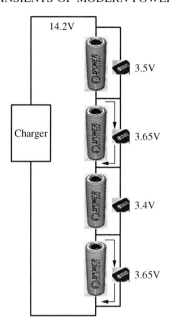

14.2V

Charger

3.5V

3.65V

3.4V

3.65V

Figure 6.14   ASIC-based charge balancing method.

6.3.7   DC–DC converter-based balancing

A DC–DC converter-based balancing circuit is shown in Figure 6.15 [15]. The battery pack is divided into two strings with each string containing the same number of cells. There is an isolated bidirectional DC–DC converter between the two strings of cells. There are two pairs of balancing buses, and 110 switches for this example of a 100-cell battery pack. The batteries and switches are divided into two parts, A and B. A is on the left side and B is on the right. Every cell can be connected to the balancing buses through the switches named KA1, KA2, . . . , KA51, KB1, KB2, . . . , KB51. But no more than two cells can be connected to the balancing buses on the same side at any given time, otherwise this will cause a short circuit because of the main serial buses.

The balancing scheme can be described by the following steps:

1. Select the highest cell in part A and the lowest cell in part B.

2. Turn on the corresponding polar switches and control the energy flow using the DC–DC converter to flow from part A to part B, if the highest cell in part A is higher than the lowest one in part B.

3. Select the lowest cell in part A and the highest cell in part B.

4. Turn on the corresponding polar switches and control the energy of the DC–DC converter to flow from part B to part A, if the highest cell in part B is higher than the lowest one in part A.

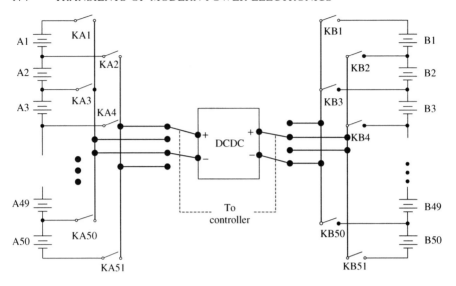

Figure 6.15 DC–DC converter-based balancing method. © [2009] IEEE. Reprinted, with permission, from VPPC'09.

5. Repeat steps 1–4 repeatedly until the highest cell voltage in A is equal to the lowest cell voltage in B and the lowest cell voltage in A is equal to the highest cell voltage in B. Then balance is reached.

Because cost and reliability are mainly determined by the number of switches, it is important to balance dozens of batteries with as simple a switching topology as possible. Dimidiation of the balancing buses is the most effective way to reduce the number of switches. From the above analysis, the polarity of balancing buses is not fixed and can be positive or negative, which is determined by which bus is connected to the DC–DC converter's positive and negative input or output. This flexibility of the balancing bus polarity reduces the number of switches by half, though it will increase the complexity of the balancing procedure. In this strategy a bidirectional DC–DC converter is the core.

A variation of the DC–DC-based converter combines the advantage of a DC–DC converter and the inductive balancing scheme as shown in Figure 6.16. In this circuit, the input of the DC–DC converter is connected to the whole string while the output is connected to each battery cell through the selective switch. During charge, discharge, or idle time, when the DC–DC converter is activated it will search for the cell with the lowest voltage among all cells. Once the cell with lowest voltage is found, it will then charge this cell using the energy from the whole string, until the cell reaches the average voltage. This process will continue until all cells have the same voltage.

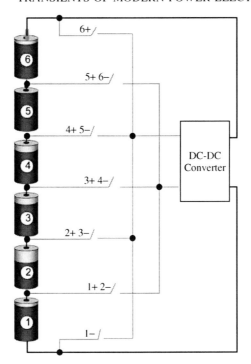

Figure 6.16   Another DC–DC converter-based balancing method.

6.4   SOA of battery power electronics

Safety is a major criterion for battery usage. Recent accidents in the consumer electronics industry and some after-market PHEVs have illustrated the importance of safety of batteries and the associated control [16–18]. Battery safety can be managed and enhanced through the use of power electronics.

6.4.1   Enhanced system-level SOA considering the battery impedance and temperature

In Chapter 4, a system-level SOA is given as a reference to charge the battery with maximized power capability. When switch temperature and battery impedance are fully considered, this SSOA is expected to be further enhanced.

First, temperature is a major concern but has mostly been ignored in the previous literature. Temperature deeply influences the switching characteristics. The thermal characteristics of the silicon chip used in Chapter 4 are: the maximum repetitive turn-off current is 70 A at $T_j = 25\,°C$, 44 A at $T_j = 100\,°C$, and 30 A at $T_j = 125\,°C$. Therefore the power capability of the whole charging system is expected to be significantly reduced at high-temperature conditions. After laying out the MOSFETs on the heat sink and obtaining the loss data based on the system model, the simulated maximum heat sink temperature is 43.75 °C at 25 °C ambient temperature, as shown in Figure 6.17. The average switch loss is 38.5 W based on the MOSFET model. According to the datasheet, the real junction temperature can be estimated as $T_j = 0.2 \times 38.5 + 43.75 = 51.45\,(°C)$, which allows 55 A maximum turn-off current.

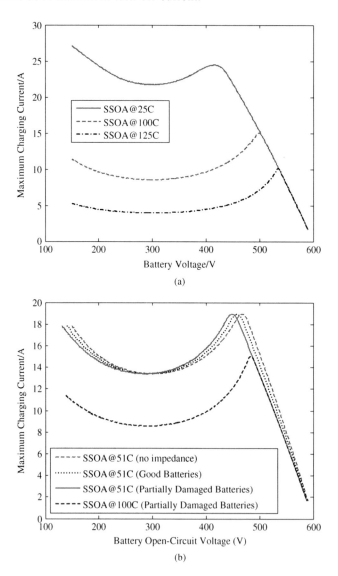

(a)

(b)

Figure 6.17 Enhanced SSOA with (a) thermal characteristics only and (b) internal impedance of the battery at different temperatures.

However, given the fact that a vehicle can operate at both very high and low environmental temperature conditions, the above estimation of maximum allowed current will need to be further adjusted. For example, in some states, the environmental temperature could be as high as 55 °C. This will put the junction temperature in the 80–90 °C temperature range, which will further decrease the charging capability of the charger.

Load characteristics are rarely included in past discussions and are believed to affect the SSOA. For the battery-charging system, the DC internal resistance varies with the battery terminal voltage at different charging current. For a single 40 Ah, 13.6 V lithium-ion battery module, the nominal maximum DC impedance is 15 mΩ. For a 365 V battery string (27 cells connected in series), the overall resistance could reach 0.5 Ω, which assumes that the battery is not aged. When some of the cells in the battery string become aged, the resistance changes significantly.

The SSOA of the charger can be further modified as shown in Figure 6.17. In Figure 6.17a, high ambient temperature sharply decreases the maximum charging capability, which directly reduces the SSOA of the battery charger. In Figure 6.17b, the battery impedance alters the shape of the SSOA. A partially aged or damaged battery will allow much lower charging current due to its high internal resistance. For an ideal battery without impedance, at an open-circuit voltage of 500 V, the maximum charging current is 15 A. However, only 13 A is allowed for the partially aged battery string.

6.4.2   Interaction with other devices at different temperatures

As a promising alternative of silicon semiconductor devices, silicon carbide (SiC) could also be used in future battery chargers. However, even for SiC devices, the influence of the dynamic transient processes and temperature on the SSOA is still significant.

The turn-on waveform of a SiC JFET in the phase without an external SiC Schottky diode is shown in Figure 6.18a. Because the body diode of the SiC JFET is a normal PN junction which contains no Schottky barrier, it has obvious reverse recovery and appears as a big current overshoot. This phenomenon is even worse at high temperature since the body diode's performance becomes worse at such a temperature. At 25 °C, the current overshoot is less than 10 A above the steady state current (15 A), but at 200 °C this overshoot increases to 20 A. When an external SiC Schottky diode is put in parallel with the SiC JFET, the reverse recovery is significantly reduced, especially at high temperature, as shown in Figure 6.18b. The current overshoot is reduced to less than 10 A above the steady state value (15 A) for 25, 100, and 200 °C, and does not change very much at different temperatures.

When the switches are equipped in the real system with other components and associated control algorithms, their interactions need to be considered. Therefore, for the SSOA, modeling and testing one single semiconductor device is not sufficient. Figure 6.18 is a typical interaction between the SiC JFET and the

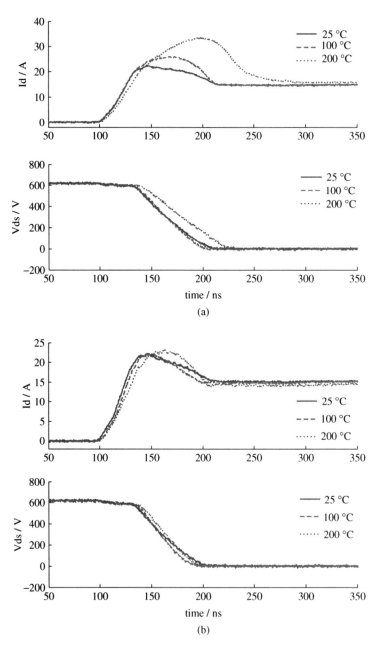

Figure 6.18   Experimental results: the turn-on waveform of SiC JFET in the phase leg (a) without external SiC Schottky diode and (b) with external SiC Schottky diode. Courtesy Professor Fred Wang, University of Tennessee.

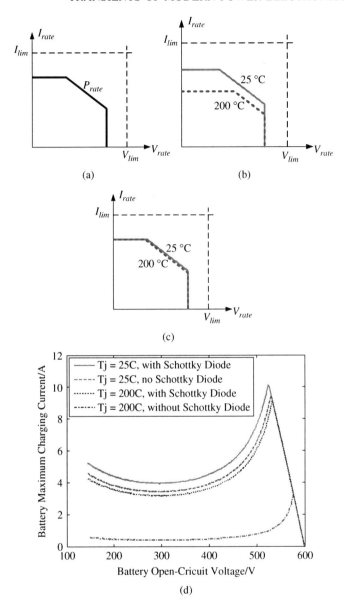

Figure 6.19   System-level SOA for the SiC JFET: (a) typical SSOA of a bridge, (b) SOA of the bridge with a normal diode as freewheeling diode, (c) SOA of the bridge with a SiC Schottky diode as freewheeling diode, and (d) SSOA for the electric charger using SiC JFETs. For Figures a-c: Courtesy Professor Fred Wang, University of Tennessee.

complementary diode. The poor diode performance will endanger the SiC JFET thereby reducing the available SSOA.

Comparing Figure 6.19a to b, the SSOA will be different for the case without an external SiC Schottky diode and the case with an external SiC Schottky diode. Up to now, there has not been sufficient data available for the SiC JFET. Therefore, the SSOA construction is very dependent on experimental data. Assuming the JFET instantaneous current limit is I_{lim} (for a 15 A SiC JFET, the maximum turn-on current is set as 25 A and the maximum turn-off current is 15 A) and the device voltage limit is V_{lim}, a typical SSOA is as shown in Figure 6.19a. The SSOA is limited by the voltage stress, current stress, and power rating. The horizontal axis is the DC-bus voltage and the vertical axis is the output peak current of the bridge. Without the SiC Schottky diode, because of the higher overshoot at higher temperature (Figure 6.18a) the bridge output current rating needs to be reduced to keep the current instantaneous value below I_{lim}, shown in Figure 6.19b. With the SiC Schottky diode, the overshoot at higher temperature does not obviously increase, as shown in Figure 6.18b. Therefore, the current rating can stay the same and the power rating will not change much either, as shown in Figure 6.19c.

Figure 6.19a–c could also be regarded as the SSOA of a two-level DC–AC inverter using SiC JFETs. With the phenomena observed in Figures 6.18 and 6.19a–c, the SSOA of a battery charger which employs SiC JFETs is shown in Figure 6.19d. The SSOA of the charger with Schottky diodes is much broader than that without Schottky diodes, especially at higher temperature. Figure 6.19d includes the temperature and battery impedance. When the junction temperature reaches 200 °C, the charger without the Schottky diodes nearly loses the charging capability.

References

1. Scholer, R.A., Maitra, A., Ornelas, E. *et al.* (2010) Communication between Plug-in Vehicles and the Utility Grid. SAE Paper Number 2010-01-0837, doi: 10.4271/2010-01-0837. http://papers.sae.org/2010-01-0837/.

2. Wikipedia (2011) http://en.wikipedia.org/wiki/Nickel-cadmium_battery (accessed March 25, 2011).

3. Meshri, D.T., Meshri, S.D., Davis, R. *et al.* (2002) Commercial scale preparation, properties and the performance of LiAsF6, LiPF6, LiBF4 electrolytes in secondary lithium ion and lithium cells. The Seventh Battery Conference on Applications and Advances, pp. 151–163.

4. Robillard, C., Vallee, A., and Wilkinson, H. (2004) The impact of lithium-metal-polymer battery characteristics on telecom power system design, 26th Annual International Telecommunications Energy Conference, pp. 25–31.

5. Nookala, M., Scanlon, L.G., and Marsh, R.A. (1997) Preparation and characterization of a hybrid solid polymer electrolyte consisting of poly(ethyleneoxide) and poly(acrylonitrile) for polymer-battery application. Proceedings of the 32nd Intersociety Energy Conversion Engineering Conference, pp. 13–18.

6. Zhang, J., Yu, J., Cha, C., and Yang, H. (2004) The effects of pulse charging on inner pressure and cycling characteristics of sealed Ni/MH batteries. *Journal of Power Sources*, **136**, 180–185.

7. NC2000 Charge Controller for Nickel-Cadmium and Nickel-Metal Hydride Batteries. User Manual, http://www.harald-sattler.de/files/NC2000_O_charger.pdf (accessed March 25, 2011).

8. Hua, C.-C. and Lin, M.-Y. (2000) A study of charging control of lead-acid battery for electric vehicles. Proceedings of the 2000 IEEE International Symposium on Industrial Electronics, Vol. 1, pp. 135–140.

9. Lin, C.-H., Hsieh, C.-Y., and Chen, K.-H. (2010) A Li-ion battery charger with smooth control circuit and built-in resistance compensator for achieving stable and fast charging. *IEEE Transactions on Circuits and Systems*, **57** (2), 506–517.

10. Bhatt, M., Hurley, W.G., and Wolfle, W.H. (2005) A new approach to intermittent charging of valve-regulated lead-acid batteries in standby applications. *IEEE Transactions on Industrial Electronics*, **52** (5), 1337–1342.

11. Li, J., Murphy, E., Winnick, J., and Kohl, P.A. (2001) The effects of pulse charging on cycling characteristics of commercial lithium-ion batteries. *Journal of Power Sources*, **102** (1), 302–309.

12. Tesla Motors (2006) The Tesla Roadster Battery System, http://webarchive.teslamotors.com/display_data/TeslaRoadsterBatterySystem.pdf (accessed March 25, 2011).

13. Moore, S.W. and Schneider, P.J. (2001) A Review of Cell Equalization Methods for Lithium Ion and Lithium Polymer Battery Systems. SAE Paper Number 2001-01-0959, doi: 10.4271/2001-01-0959.

14. Mi, C. (2008) ASIC based active charger balancing lithium battery to extend life and maximize capacity of mobile and laptop batteries. Annual International Conference on Lithium Mobile Power.

15. Nie, Z. and Mi, C. (2009) Fast battery equalization with isolated bidirectional DC–DC converter for PHEV applications. Vehicle Power and Propulsion Conference, pp. 78–81.

16. NRECA and US Department of Energy (2008) Report of Investigation: Hybrids Plus Plug In Hybrid Electric Vehicle, http://www.evworld.com/library/prius_fire_forensics.pdf (accessed March 25, 2011).

17. US Consumer Product Safety Commission (2008) PC Notebook Computer Batteries Recalled Due to Fire and Burn Hazard, http://www.cpsc.gov/cpscpub/prerel/prhtml09/09035.html (accessed March 25, 2011).

18. US Consumer Product Safety Commission (2006) Sony Recalls Notebook Computer Batteries Due To Previous Fires, http://www.cpsc.gov/cpscpub/prerel/prhtml07/07011.html (accessed March 25, 2011).

7

Dead-band effect and minimum pulse width

The topology of a typical three-level DC–AC inverter based on IGCTs is shown in Figure 7.1a. For the switches, such as Sa1, Sa2, Sa3, and Sa4, two influential short-timescale factors need to be considered. One is the dead band, and the other is the minimum pulse width (MPW):

1. **Dead band:** a dead band is inserted between the interlocked switches to guarantee that the switches are never turned on simultaneously, thereby preventing shoot-through of the DC bus [1, 2]. In Figure 7.1, Sa1 and Sa3 are the interlocked switches which cannot be turned on simultaneously when Sa2 is on. Otherwise, the upper voltage source will be shorted and a large in-rush current will occur. The ideal pulse pair is shown in Figure 7.1b. In order to avoid shoot-through, a time interval, T_{db}, is inserted between the gate signal sequence shown in Figure 7.1c.

2. **MPW:** MPV is another constraint for the system application [3, 4]. Dead band is applicable to two or more interlocked switches while MPW is for a single switch. A semiconductor switch needs some specific interval to turn on/off effectively. Therefore the pulse width calculated by the ideal control algorithm may be too narrow to fully switch on/off the semiconductors as shown in Figure 7.1d.

The width of narrow pulse 1 is less than T_{onm}, the minimum turn-on pulse width, and the width of narrow pulse 2 is less than T_{offm}, the minimum turn-off pulse width. After modifying the control algorithm, these pulses are adjusted to the required MPW. Those that are larger than MPW remain the same.

Transients of Modern Power Electronics, First Edition. Hua Bai and Chris Mi.
© 2011 John Wiley & Sons, Ltd. Published 2011 by John Wiley & Sons, Ltd.

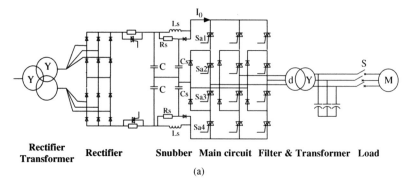

(a)

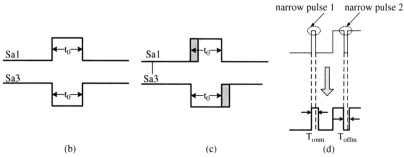

(b)   (c)   (d)

Figure 7.1   Dead band and minimum pulse: (a) a three-level inverter, (b) pulse pairs without dead band, (c) pulse pairs with dead band, and (d) minimum pulse width.

To achieve safe and reliable operation, dead band and MPW are critical requirements. However, the existence of these two parameters distorts the gate signals from the ideal ones, as shown in the following sections of this chapter. In this chapter, we will take a three-level DC–AC inverter and a DC–DC converter as examples to illustrate the effect of dead band and MPW. We address their influences on the macro-control algorithms, and propose strategies to mitigate their negative impact.

7.1   Dead-band effect in DC–AC inverters

In the dead band, all the interlocked switches are off. Hence the system is in an uncontrolled state. Some abnormal phenomena will emerge, for example, distorting the output voltage of the DC–AC voltage source inverter. In this section, we will discuss the dead-band effect based on a space vector control strategy [5].

Consider Sa1 and Sa3 in Figure 7.1a as an example. We define "1" as the on state and "0" as the off state of a switch. In one bridge, normally we can define three states of output voltage: **P**, **0**, and **N**. **P** means the output voltage of the bridge is one-half of the bus voltage, $+V_{dc}/2$. Therefore, in the **P** state, Sa1 and Sa2 are on while Sa3 and Sa4 are off. Hence state **P** corresponds to [Sa1, Sa2, Sa3, Sa4] = [**1, 1, 0, 0**]. Similarly **0** equals [**0, 1, 1, 0**] where Sa2 and Sa3 are on while Sa1 and Sa4 are off, thereby outputting zero voltage. In the **N** state, the output voltage is negative one-half of the bus voltage, $-V_{dc}/2$, where Sa3 and Sa4 are on while Sa1 and Sa2 are off. So **N** stands for the state [Sa1, Sa2, Sa3, Sa4] = [**0, 0, 1, 1**].

Suppose the state changes from **P** to **0**, that is, [**1, 1, 0, 0**]→[**0, 1, 1, 0**]. Here Sa1 turns off and Sa3 turns on at the same time with the states of other two switches unchanged. If switching on Sa3 is prior to the action of Sa1, that is, there is an overlap when Sa1 and Sa3 are both on, then an unexpected region [Sa1, Sa2, Sa3, Sa4] = [**1, 1, 1, 0**] appears, which will short the upper DC bus. This could happen if there is no dead band between the two gate signals.

It is therefore necessary to insert one dead band in the pulse sequence to eliminate the state of [**1, 1, 1, 0**] effectively. Additional states will appear due to dead band. When the state changes from **P** to **0**, that is, [**1, 1, 0, 0**] to [**0, 1, 1, 0**], state D_1 = [**0, 1, 0, 0**] with the duration of one dead-band time is inserted. Technically, Sa1 is first turned off and then, after one dead band, Sa3 is turned on. The real-time sequence is **P**←→D_1←→**0**. Likewise, when the state jumps from **0** to **N**, D_2 = [**0, 0, 1, 0**] is inserted and therefore the real sequence is **0**←→D_2←→**N**.

In state D_1, [**0, 1, 0, 0**], if the current direction is from the inverter bridge to the motor, the output voltage is **0**. Otherwise, the output voltage is **P** where current is freewheeling through the anti-parallel diodes of Sa1 and Sa2. In other words, the output voltage is determined by the current direction in this uncontrolled state. This is defined as dead-band effect.

This dead-band effect influences the output voltage directly, so normally we can detect this phenomenon by measuring the relevant voltages, for example, the line-to-line voltage. In the following, we will take one control algorithm, seven-section space vector PWM, as an example.

Example 7.1   Seven-Section Space Vector Control

From the above analysis, the output voltage of each bridge has three states, **P**, **0**, **N**, that is, $+V_{dc}/2$, 0, and $-V_{dc}/2$. Therefore, for the three phases, there are 3^3 = 27 voltage vectors available. Figure 4.14a in Chapter 4 shows the vector distribution given for the three-level inverter. For example, $\vec{V}_1 = PNN$ means that the output voltages of phase A, B, and C are $+V_{dc}/2$, $-V_{dc}/2$, and $-V_{dc}/2$, respectively.

The actual voltage vector imposed on the load $\vec{V}_{ref}$ is in fact an equivalent combination of the three adjacent vectors in the same sector of the figure. For example, $\vec{V}_{ref}$ comprises $\vec{V}_{02}$, $\vec{V}_{01}$, and $\vec{V}_{12}$:

$$\vec{V}_{ref} \times T_s = \vec{V}_{01} \times T_1 + \vec{V}_{12} \times T_2 + \vec{V}_{02} \times T_3$$

$$T_s = T_1 + T_2 + T_3 \tag{7.1}$$

In order to make the voltage transition smooth and to utilize the advantage of three-level topology, no voltage leap from **P** to **N** or **N** to **P** for each phase is allowed. Hence the vector sequence should be

$$\textbf{P00} \rightarrow \textbf{P0N} \rightarrow \textbf{00N} \rightarrow \textbf{0NN} \rightarrow \textbf{00N} \rightarrow \textbf{P0N} \rightarrow \textbf{P00} \tag{7.2}$$

Then the output voltages of phase A, B, and C are

U_A: **P** → **P** → **0** → **0** → **0** → **P** → **P**

U_B: **0** → **0** → **0** → **N** → **0** → **0** → **0**

U_C: **0** → **N** → **N** → **N** → **N** → **N** → **0**

t: $T_1/4 \rightarrow T_2/2 \rightarrow T_3/2 \rightarrow T_1/2 \rightarrow T_3/2 \rightarrow T_2/2 \rightarrow T_1/4$ (7.3)

Therefore the line-to-line voltage $U_{AB} = U_A - U_B$ is

U_{AB}: **P** → **P** → **0** → **P** → **0** → **P** → **P**

7.1.1 Dead-band effect

Equation 7.3 stands for ideal operation modes without dead band, as shown in Figure 7.2a. In the motor drive, for the convenience of analysis, the current is assumed unchanged within one switching period T_s. When U_A changes from **P** → **0**, the time of **00N**, $T_3/2$, will be reduced by T_{db} when current I_A is from inverter to motor, as shown in Figure 7.2b. Similar analysis can be applied to phase B. If T_3 is small enough, the duration of **00N** will be totally ignored and therefore Figure 7.2b will be changed to Figure 7.2c, where two abnormal pulses appear. If the current changes its direction within one switching period, the output voltage can even be distorted as shown in Figure 7.2d, which has been validated experimentally as shown in Figure 7.2e [6, 7].

This dead-band effect is visible in the DC–AC inverter and behaves differently with the control algorithms [8]. It is worthwhile to point out that, although it distorts the voltage waveform, this distortion does little harm to system reliability.

In special cases, if we assume the current direction does not change, the distortion of voltage or current waveforms due to dead band can be calculated quantitatively.

Example 7.2   Vectors of **0NP** and **000** are applied in turn to control the direct current injected into the motor to establish the electromagnetic field during pre-excitation of the motor. Ideally, current I_C is from bridge to motor, I_B is in the opposite direction, and $I_A = 0$. A comparison of the output voltage waveforms

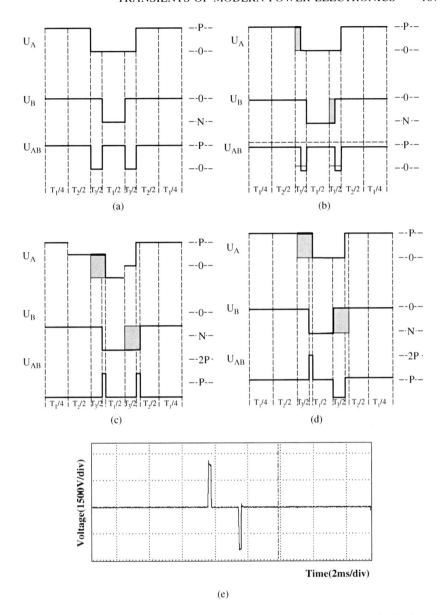

Figure 7.2 Dead-band effect: (a) ideal waveform without dead band; (b) with $T_{db} < T_3/2$; (c) with $T_{db} > T_3/2$; (d) asymmetric dead-band effect when $T_{db} > T_3/2$; and (e) experimental line-to-line voltage ($V_{DC} = 3700\,V$, $P = 1.25\,MW$).

with/without dead band is shown in Figure 7.3a where the light line corresponds to the ideal waveforms and the bold line corresponds to the real ones.

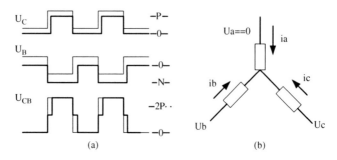

(a)   (b)

Figure 7.3   The dead-band effect in the pre-excitation of motor drive: (a) 0NP/000, (b) scheme of Y-connected resistors.

Suppose the impedance of each phase is R (since all the current and voltage components are DC components, the equivalent inductance of the motor at steady state is neglected). According to superposition theory,

$$i_A = \frac{1}{3R}(U_C + U_B)$$

$$i_B = \frac{2U_B}{3R} - \frac{U_C}{3R}$$

$$i_C = \frac{2U_C}{3R} - \frac{U_B}{3R} \tag{7.4}$$

From Figure 7.3, the dead-band effect changes the waveforms of the phase voltage and line-to-line voltage. In more detail,

$$i'_A = \frac{1}{3R}\left(U^*_C - \frac{T_{db}}{T_s}\frac{V_{dc}}{2} + U^*_B - \frac{T_{db}}{T_s}\frac{V_{dc}}{2}\right) = i^*_A - \frac{V_{dc}}{3R}\frac{T_{db}}{T_s} = -\frac{V_{dc}}{3R}\frac{T_{db}}{T_s}$$

$$i'_B = \frac{2U_B}{3R} - \frac{U_C}{3R} = \frac{2\left(U^*_B - \frac{T_{db}}{T_s}\frac{V_{dc}}{2}\right)}{3R} - \frac{\left(U^*_C - \frac{T_{db}}{T_s}\frac{V_{dc}}{2}\right)}{3R} = i^*_B - \frac{V_{dc}}{6R}\frac{T_{db}}{T_s}$$

$$i'_C = \frac{2U_C}{3R} - \frac{U_B}{3R} = \frac{2\left(U^*_C - \frac{T_{db}}{T_s}\frac{V_{dc}}{2}\right)}{3R} - \frac{\left(U^*_B - \frac{T_{db}}{T_s}\frac{V_{dc}}{2}\right)}{3R} = i^*_C - \frac{V_{dc}}{6R}\frac{T_{db}}{T_s} \tag{7.5}$$

Therefore, the dead-band effect changes the phase current used to establish the electromagnetic field during pre-excitation. In most cases, the influence of dead band cannot be quantitatively analyzed as Equation 7.5 because of the nonlinear relation between voltage and current.

7.2   Dead-band effect in DC–DC converters*

In some DC–DC converters, such as those based on the dual active bridge (DAB), similar phenomena can be observed [8].

7.2.1   Phase shift-based dual active bridge bidirectional DC–DC converter

This DC–DC converter consists of two H-bridges located on the primary and secondary sides of an isolation transformer. The primary bridge, assembled by four switches, Q_1, Q_2, Q_3, and Q_4, is connected to the primary winding of the isolation transformer. The secondary H-bridge, also assembled by four switches, Q_5, Q_6, Q_7, and Q_8, is connected to the secondary winding of the transformer. For the phase-shift control algorithm, the first H-bridge provides a square wave with a duty ratio of 50% to the primary winding of the high-frequency transformer. The voltage of the secondary winding of the transformer has a definite phase-shift angle from the primary voltage so as to transfer energy from one side to the other. In this process the transformer's leakage inductor serves as the instantaneous energy storage component.

In the following analysis, the turns ratio of the transformer is n, the transformer's primary voltage is V_1 and the switching frequency is f_s. For the convenience of analysis, we define one half period as T_s, that is, $T_s = 1/(2f_s)$. The duty ratio or phase shift is based on the half period, $D = t_{on}/T_s$. Therefore, DT_s is the phase shift between the two bridges. Further, I_{L_s} is the current of the leakage inductance of the secondary winding. The output voltage of the secondary bridge is V_2. We assume $V_2 > nV_1$.

First, we neglect the dead-band effect, which will be analyzed in the latter part of this section. There are a number of different operation modes based on the output current, with a boundary condition, as illustrated in Figure 7.4. Figure 7.4a corresponds to heavy load conditions where the inductor current increases from an initial negative value $i(t_0) < 0$ at the beginning of the switching cycle, and reaches a positive value $-i(t_0)$ at the end of the half switching cycle. Six different segments emerge in each switching cycle as shown in Figure 7.4. The detailed switch operations are illustrated in Figure 7.5 [9]:

Segment 0, $[t_0,t_1]$: in this segment, Q_1 and Q_4 of the primary bridge are turned on. Therefore V_1 and nV_1 are positive. Q_6 and Q_7 of the secondary bridge are turned on. Due to the negative current in the inductor, D_6 and D_7 freewheel, and Q_6 and Q_7 do not conduct current: $V_{L_s} = nV_1 + V_2$. The inductor current increases linearly from a negative value. At t_1, the inductor current reaches 0.

Segment 1, $[t_1,t_2]$: switches Q_1 and Q_4 of the primary bridge, and Q_6 and Q_7 of the second bridge, are still turned on: $V_{L_s} = nV_1 + V_2$. The current

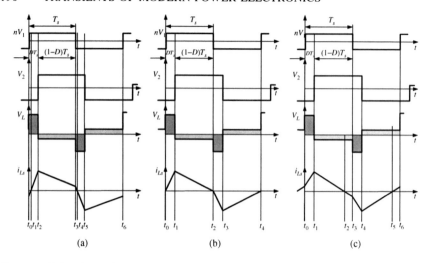

Figure 7.4 Typical voltage and current waveforms $(D > 0,\ V_2 > nV_1)$: (a) $i(t_0) < 0$, (b) boundary conditions $i(t_0) = 0$, (c) $i(t_0) > 0$.

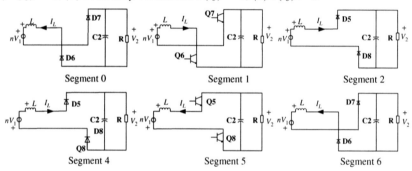

Figure 7.5   The switch modes relating to Figure 7.4.

continues to increase except that it becomes positive, that is, Q_6 and Q_7 conduct the current. The total current increment during interval $[t_0, t_2]$ (**segments 0** and **1**) is

$$\Delta I_{L_s} = \frac{DT_s}{L_s}(V_2 + nV_1) \tag{7.6}$$

Hence

$$i(t_2) = i(t_0) + \frac{DT_s}{L_s}(nV_1 + V_2) = I_{max} \tag{7.7}$$

Segment 2, $[t_2,t_3]$: in this segment, switches Q_1 and Q_4 of the primary bridge continue to be turned on but switches Q_6 and Q_7 are turned off, and switches Q_5 and Q_8 are turned on. Diodes D_5 and D_8 freewheel because the current is positive: $V_{L_s} = nV_1 - V_2 < 0$.

The leakage inductor current increment during interval $[t_2, t_3]$ is

$$\Delta I_{L_s} = \frac{(1 - D)T_s}{L_s}(nV_1 - V_2) \tag{7.8}$$

Therefore

$$i(t_3) = i(t_2) + \frac{(1 - D)T_s}{L_s}(nV_1 - V_2) \tag{7.9}$$

Similar analyses could be pursued for the following three segments.

Segment 4, $[t_3, t_4]$: switches Q_2 and Q_3 of the primary bridge continue to be turned on, and switches Q_5 and Q_8 are turned on. The primary and voltage of the transformer is reversed and the current decreases from $i(t_3)$ to zero. D_5 and D_8 freewheel.

Segment 5, $[t_4, t_5]$: the switch mode is the same as **segment 4** except that the current decreases linearly to the negative maximum. Switches Q_5 and Q_8 conduct current. Hence the current increment in L_s in **segments 4** and **5** is

$$\Delta I_{L_s} = -\frac{DT_s}{L_s}(nV_1 + V_2) \tag{7.10}$$

Segment 6, $[t_5, t_6]$: Q_5 and Q_8 are turned off, and D_6 and D_7 begin to freewheel. The current increment in L_s is

$$\Delta I_{L_s} = \frac{(1 - D)T_s}{L_s}(V_2 - nV_1) \tag{7.11}$$

In summary, the above seven segments can be categorized into four different state equations corresponding to the different time intervals. Assuming V_1 is constant, then

$$\begin{bmatrix} \dfrac{di_L}{dt} \\[2ex] \dfrac{dV_2}{dt} \end{bmatrix} = \begin{bmatrix} 0 & \dfrac{1}{L} \\[2ex] -\dfrac{1}{C_2} & -\dfrac{1}{RC_2} \end{bmatrix} \begin{bmatrix} i_L \\[2ex] V_2 \end{bmatrix} + \begin{bmatrix} \dfrac{1}{L} \\[2ex] 0 \end{bmatrix} nV_1$$
$$t \in [0 \quad DT_s] \tag{7.12}$$

$$\begin{bmatrix} \dfrac{di_L}{dt} \\[2ex] \dfrac{dV_2}{dt} \end{bmatrix} = \begin{bmatrix} 0 & -\dfrac{1}{L} \\[2ex] \dfrac{1}{C_2} & -\dfrac{1}{RC_2} \end{bmatrix} \begin{bmatrix} i_L \\[2ex] V_2 \end{bmatrix} + \begin{bmatrix} \dfrac{1}{L} \\[2ex] 0 \end{bmatrix} nV_1$$
$$t \in \begin{bmatrix} DT_s & T_s \end{bmatrix} \tag{7.13}$$

$$\begin{bmatrix} \dfrac{di_L}{dt} \\[2ex] \dfrac{dV_2}{dt} \end{bmatrix} = \begin{bmatrix} 0 & -\dfrac{1}{L} \\[2ex] \dfrac{1}{C_2} & -\dfrac{1}{RC_2} \end{bmatrix} \begin{bmatrix} i_L \\[2ex] V_2 \end{bmatrix} + \begin{bmatrix} -\dfrac{1}{L} \\[2ex] 0 \end{bmatrix} nV_1$$

$$t \in [T_s \ (1+D) \, T_s] \tag{7.14}$$

$$\begin{bmatrix} \dfrac{di_L}{dt} \\[2ex] \dfrac{dV_2}{dt} \end{bmatrix} = \begin{bmatrix} 0 & \dfrac{1}{L} \\[2ex] -\dfrac{1}{C_2} & -\dfrac{1}{RC_2} \end{bmatrix} \begin{bmatrix} i_L \\[2ex] V_2 \end{bmatrix} + \begin{bmatrix} -\dfrac{1}{L} \\[2ex] 0 \end{bmatrix} nV_1$$

$$t \in [(1+D) \, T_s \ 2T_s] \tag{7.15}$$

because of the symmetry of the inductor current, $i(t_0) = -i(t_3)$. From Equations 7.6–7.9, the initial inductor current can be obtained as

$$i(t_0) = \frac{1}{4 f_s L_s} [(1 - 2D)V_2 - nV_1] \tag{7.16}$$

The maximum current is

$$I_{\max} = \frac{1}{4 f_s L_s} [-(1 - 2D)nV_1 + V_2] \tag{7.17}$$

The above analysis of operation modes is based on the assumption that $i(t_0) < 0$, that is, $(1 - 2D)V_2 < nV_1$. If $(1 - 2D)V_2 = nV_1$ or

$$V_2 = \frac{1}{1 - 2D} nV_1 \tag{7.18}$$

then $i(t_0) = 0$. This corresponds to the boundary condition as shown in Figure 7.4b which is very similar to the boundary condition of a non-isolated boost converter in steady state operation. At this condition the inductor current increases from zero at the beginning of the switching cycle, and drops to zero at the end of T_s.

Thus we can start with the average current in the inductor to calculate the average power. The average current in the leakage inductance is

$$\bar{I} = \frac{1}{2T_s} [(i_{\max} + i(t_0))DT_s + (i_{\max} - i(t_0))(1 - D)T_s] = \frac{1}{2 f_s L_s} D(1 - D)V_2 \tag{7.19}$$

Therefore the supplied power is

$$P_1 = nV_1 \bar{I} = \frac{nV_1 V_2}{2 f_s L_s} D(1 - D) \tag{7.20}$$

When $D > 0$, energy flows from the primary side to the secondary side. Otherwise, energy flows from the secondary side to the primary side.

7.2.2   Dead-band effect in DAB bidirectional DC–DC converter

The above analysis addresses macroscopic energy flow based on the steady state analysis and Equation 7.20 originates from steady state operations. When the dead band is in place, the operation of the DAB-based DC–DC converter will vary.

7.2.2.1   Abnormal voltage pulse

In the dead band, all four semiconductors in the primary H-bridge, Q_1–Q_4, will be turned off. In this interval, if the inductor current crosses zero, it will cause the polarity of the output voltage to change, similar to the analysis of the DC–AC inverter. However, in the DC–AC motor drive system, load characteristics determine that the current can be regarded as a constant DC component in one switching period. In the DAB DC–DC converter, current oscillates at high frequency, which increases the possibility of oscillation of the primary voltage as shown in Figure 7.6a. The simulation shown in Figure 7.6b validates this analysis. The oscillations in the primary voltage are caused by the change of current direction in the dead band. These oscillations of voltage will not significantly deteriorate system operation, but sometimes frequent variations of voltage waveforms may generate EMI and reduce the effective phase shift.

7.2.2.2   Reduction of phase-shift angle

The dead band will affect not only the dynamic commutating process, but also the steady state performance. We define the phase shift Φ^* as the angle difference between the gate signals of Q_1 and Q_5 without the dead band, and Φ as that with the dead band. Figure 7.7 shows a comparison of the phase shift between the gate signals of Q_1 and Q_5 with/without dead band under the same output voltage and power. It is obvious that, with the increase of load, the dead-band effect is less.

   Figure 7.8 illustrates the influence of the dead band on the phase shift. Q_1–Q_8 are the gate signals on the corresponding switches. The dark area in Q_1–Q_8 is the dead band which pushes the rising edges of gate signals backward accordingly, turning off all the switches in the same H-bridge modules in the dead band.

   The output voltage in the dead band is determined by the current direction. When $i(t0) < 0$ (inside the bridge), the primary output voltage is nearly the same with the no-dead-band ideal operation. Hence $\Phi = \Phi^*$ under the same output power. When $i(t_0) > 0$, a phase shift Φ_{db} is erased from the output voltage due to the dead band. Hence for $i(t_0) > 0$, in order to maintain the same output power as the ideal operation, $\Phi = \Phi^* + \Phi_{db}$ is required where $\Phi_{db} = T_{dead\ band} \times 2\pi f_s$.

   It is worthwhile to point out that, due to the possible change of current direction in one dead band, the erased phase shift, that is, the differences of the two lines in Figure 7.7, are not exactly one dead band or zero, but a nonlinear function of output power, which will be discussed later.

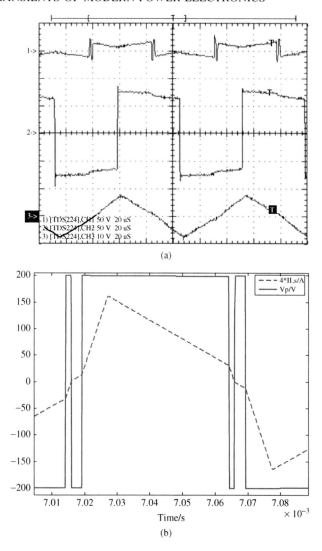

Figure 7.6   Dead-band effect on the primary voltage waveforms: (a) experiment, (b) simulation. Channel 1 → V_1, Channel 2 → V_2, Channel 3 → I_p (primary current).

7.2.2.3   Current distortion

In the dead band, the H-bridge behaves as a rectifier, where the polarity of the voltage imposed on the primary/secondary terminals depends on the direction of $i(t_0)$. At the same time, the current slope is closely related to the imposed voltage, as shown in Figure 7.9. Before current changes from negative to positive, the current slope is $(nV_1 + V_2)/L_s$; after that, it becomes $(-nV_1 + V_2)/L_s$. After the dead band, it changes back to $(nV_1 + V_2)/L_s$.

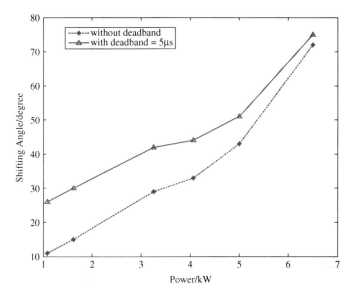

Figure 7.7   Shifted phase comparison in the steady states (simulation).

7.2.2.4   Energy dead zone

Without the dead band, the time sequence of gate signals should be as follows:

1. Q_1 and Q_4 turn on while Q_2 and Q_3 turn off simultaneously.

2. After a phase shift DT_s, Q_6 and Q_7 are turned off with Q_5 and Q_8 turned on.

Thus

$$Q_1/Q_4 \quad \text{on} \quad \xrightarrow{DT_s} \quad Q_5/Q_8 \quad \text{on}$$

$$Q_2/Q_3 \quad \text{off} \quad \rightarrow \quad Q_6/Q_7 \quad \text{off}$$

However, due to the dead band, the time sequence of each gate signal is changed to

$$Q_2/Q_3 \quad \text{off} \quad \xrightarrow{db_1} \quad Q_1/Q_4 \quad \text{on} \quad \xrightarrow{\Delta} \quad Q_6/Q_7 \quad \text{off} \quad \xrightarrow{db_2} \quad Q_5/Q_8 \quad \text{on}$$

where db_1 and db_2 are the dead bands in the primary and secondary H-bridges, respectively. First Q_2 and Q_3 turn off. After a time equal to db_1, Q_1 and Q_4 turn on. After a time Δ, Q_6 and Q_7 turn off. Eventually, after db_2, Q_5 and Q_8 turn on. Hence the time sequence with dead band is different from the ideal ones. Within every dead band db_1 or db_2, the primary or secondary bridge behaves as a rectifier. Therefore the energy flow is as shown in Figure 7.10 [10].

Since the primary side and secondary side are equipped with different switches targeting different current ratings, the dead bands of the two H-bridges may not be the same. For example, $db_1 > db_2$. If the load is very light, there will be another extreme condition where the time sequence of the gate signals changes to the following:

$$Q_2/Q_3 \quad \text{off} \quad \xrightarrow{\Delta_1} \quad Q_6/Q_7 \quad \text{off} \quad \xrightarrow{db_2} \quad Q_5/Q_8 \quad \text{on} \quad \xrightarrow{\Delta_2} \quad Q_1/Q_4 \quad \text{on}$$
$$\xrightarrow{\hspace{2cm}} db_1 \hspace{2cm}$$

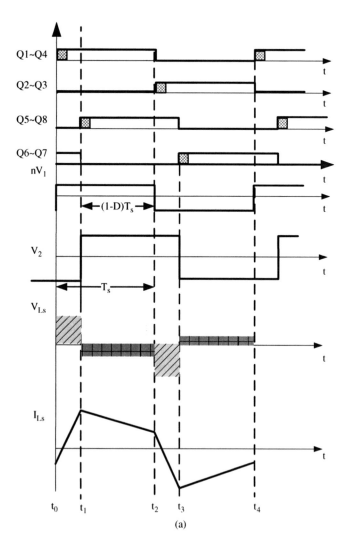

Figure 7.8   Impact of dead band on phase shift: (a) heavy load $(i(t_0) < 0)$ and (b) light load $(i(t_0) > 0)$.

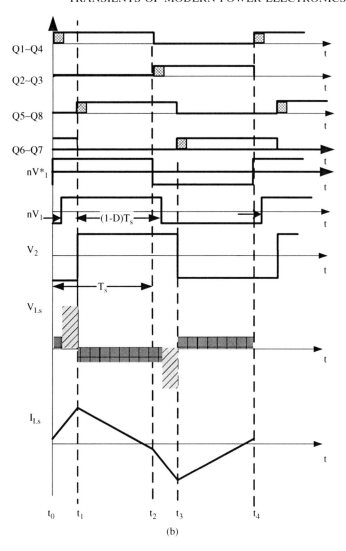

Figure 7.8   (continued)

This means that, at some conditions, turning off Q_6 could even be prior to turning on Q_5, thus the energy distribution will be changed. The time sequence of different signals is shown in Figure 7.11a. Energy flows are shown in Figure 7.11b–d.

Therefore, there is always an overlapped region where no energy flows from one side to the other but only flows from the transformer to the two DC sides, as illustrated in Figure 7.11c. In this condition, the energy stored in the leakage inductance will be consumed very quickly, and the current will remain close to zero, which causes oscillations of voltage drop on the leakage inductance as

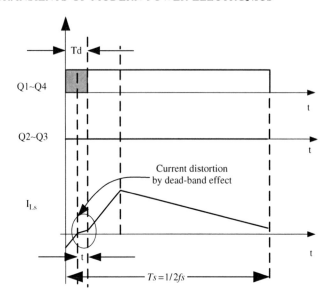

Figure 7.9   Current distortion due to dead band.

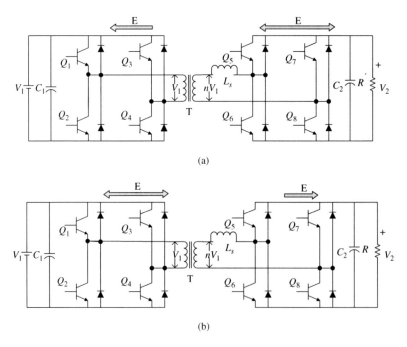

Figure 7.10   Energy flows in the different dead bands: (a) energy flows during interval db_1, (b) energy flows during interval db_2. © [2008] IEEE. Reprinted, with permission, from IEEE Transactions on Power Electronics.

shown in Figure 7.11a. This region could be defined as an *energy dead zone* since there is no energy transfer during this interval.

An energy dead zone is a special phenomenon in isolated bidirectional DC–DC converters. The condition for this to happen is light load with small phase-shift duty ratio D. Within the energy dead zone, the transformer leakage inductance behaves as the energy supply, and the original power supplies serve as energy storage. Energy in the leakage inductor will be distributed evenly to the two DC buses. The whole system is in an uncontrolled state.

7.3 Control strategy for the dead-band compensation

The dead band will affect the performance of various power converters based on the above analysis. There have been significant efforts on dead-band compensation to mitigate the dead-band effect. This includes two aspects:

To set the dead band appropriately: if a dead band is set too small, it may not prevent shoot-through of the bridge effectively. On the other hand, if a dead band is set too large, it will deteriorate the operational performance. Normally the dead-band setting is determined by the semiconductor device itself. For example, for an IGCT, there is a time interval between the edge of turn-off triggering signals and the point of 80% current value, defined as the turn-off delay time t_{doff}. The counterpart for the turn-on process is defined as t_{don}. When an IGCT is turned off, the voltage rises with

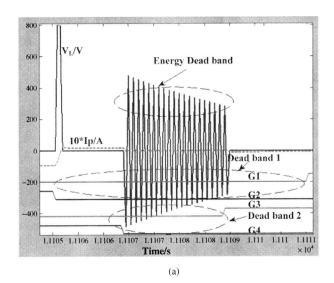

(a)

Figure 7.11 Energy flow in the very light-load condition when $db_1 > db_2$: (a) energy dead zone, (b) energy flows during interval Δ_1, (c) energy flows during interval db_2, (d) energy flows during interval Δ_2. © [2008] IEEE. Reprinted, with permission, from IEEE Transactions on Power Electronics.

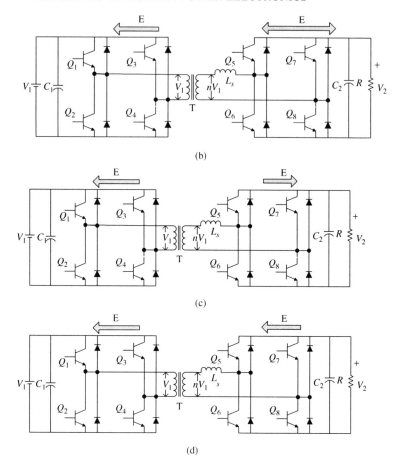

Figure 7.11   (continued)

the drop of current. However, the current does not drop to zero directly but has some tailing current lasting for a time t_{trail}, the tailing time. The interlocked switches should be turned off as the tailing current fades, as shown in Figure 7.12 [11].

Therefore the time to fully turn off the semiconductor is $t_{doff} + t_f + t_{trail}$. And the time to turn on the semiconductor is $t_{don} + t_r$. Technically the dead band for the semiconductor could be set as $(t_{doff} + t_f + t_{trail}) - t_{don}$. Considering the diversity of semiconductors, the turn-on time could be more or less. In order to provide sufficient time to turn off the switches prior to switching on the interlocked one, the dead band could be set as

$$T_{db} = t_{doff} + t_f + t_{trail} \qquad (7.21)$$

The above parameters can be found in any datasheet for the semiconductor.

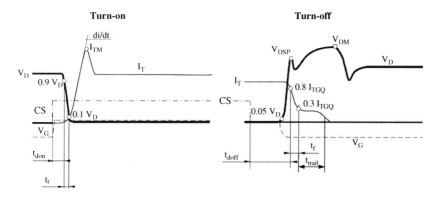

Figure 7.12   Turn-on and turn-off processes. Courtesy of ABB Switzerland Ltd, semiconductors.

To compensate the dead-band effect with macroscopic control algorithms: in the literature, dead-band compensation is mainly targeted at DC–AC inverters. A similar strategy in DC–DC converters is not yet prevalent. The main reason is that, in low-voltage and low-power systems, the dead-band setting is mostly small, which does not significantly affect the operational performance. In high-voltage and high-power systems, the dead-band setting occupies a significant portion of the phase shift, which will have a considerable impact on the system performance and needs to be studied accordingly. Here we take the dead-band compensation for the DAB-based DC–DC converter as an example.

Example 7.3   Dead-Band Compensation for DAB-Based DC–DC Converter

Compared to the dead-band effect in motor controls, the DAB-based DC–DC converter has its own characteristics. In a motor drive, the dead band will occur in both light-load and heavy-load conditions. However, in DAB-based DC–DC converters the dead band will only affect the shift angle between Q_1 and Q_5 in light-load conditions, not in heavy-load conditions. Furthermore, this erased phase-shift angle due to the dead band is a nonlinear function of load.

Suppose $m = V_2/(nV_1) > 1$; then

$$i(0) = i(t_0) = \frac{1}{4L_s f_s}[(1 - 2D)V_2 - nV_1] \tag{7.22}$$

Solving for D in Equation 7.22 and substituting into Equation 7.20, we have the relationship between power and Δt as given in Table 7.1. Here we consider the transformer scheme as in Figure 7.13a and derive the relationship between $i(0)$ and P. Thus

$$V_2' = \frac{L_m}{L_2 + L_m}V_2, \quad L_s' = L_1' + \frac{L_2 L_m}{L_2 + L_m}, \quad m' = \frac{L_m}{L_2 + L_m}\frac{V_2}{nV_1} > 1$$

Table 7.1   Relationship between power and time interval to compensate the dead band.

Power P	Erased shift time Δt
$P > \frac{nV_1V_2}{8f_sL_s}\left[1 - \left(\frac{1-4f_sT_d(1+m)}{m}\right)^2\right]$	$\Delta t = 0$
$\frac{nV_1V_2}{8f_sL_s}\left[1 - \left(\frac{1}{m}\right)^2\right] < P \leqslant \frac{nV_1V_2}{8f_sL_s}$ $\left[1 - \left(\frac{1-4f_sT_d(1+m)}{m}\right)^2\right]$	$\Delta t = T_d - \dfrac{1 - \sqrt{1 - \dfrac{8f_sL_s'P}{nV_1V_2'}m'}}{4(1+m')f_s}$
$P \leqslant \frac{nV_1V_2}{8f_sL_s}\left[1 - \left(\frac{1}{m}\right)^2\right]$	$\Delta t = T_d$

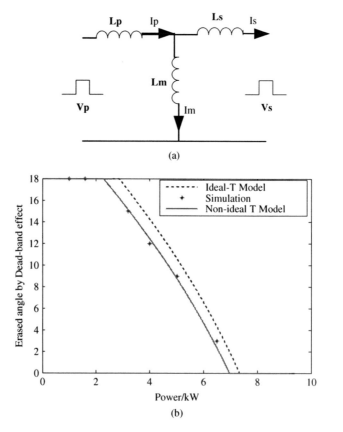

(a)

(b)

Figure 7.13   Erased angle under different models: (a) transformer model with exciting inductance and (b) adjustment of dead-band compensation.

where L_1' is the equivalent leakage inductance from the primary to the secondary side, L_2 is the secondary-side leakage inductance, and L_m is the magnetizing inductance.

The erased angles based on the ideal model of the isolated transformer, non-ideal model of the isolated transformer (with the exciting inductance), and simulations are compared in Figure 7.13b.

In Figure 7.13 the phase-shift difference Δt is exactly the same as T_d at very light load conditions, and zero at heavy load conditions. Under medium-power output, the simulation and calculation based on Table 7.1 show good agreement. It is worthwhile to point out that the exciting inductance L_m needs to be considered, especially in the light-load operation. The existence of excitation current may change the current waveform, direction, and magnitude, which is of importance in dead-band compensation.

Therefore, if we know the demanded power P, then Δt can be derived to compensate the dead band which erases portion of the calculated phase shift. In Figure 7.14, we propose a novel phase-shift controller which combines a traditional PI controller, dead-band compensator, and phase-shift predictor. Essentially this is a feedback and feedforward control, whose key point lies in the identification of load resistance. This control is to preset a value D^*, a phase-shift ratio predicted by the on-line parameter identification. However, D^* is only an ideal value without consideration of dead band T_d. A dead-band compensator is intended to further assist the PI controller. Finally, a PI controller is used to fine-tune the phase shift to satisfy the output requirement. In the light-load condition where D^* is close to zero, the output of the dead-band compensator will be dominant in determining the actual phase-shift angle. For example, when

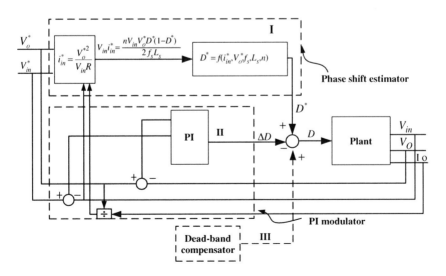

Figure 7.14   Dead-band compensator based on hybrid phase-shift controller.

$P = 1\,kW$, the calculated ideal phase shift is $11°$. Due to the existence of the dead band, the actual phase-shift angle to maintain a $1\,kW$ output is $26°$, which means nearly $15°$ will be from the PI controller and dead-band compensation.

In Figure 7.14, part I is to predict the theoretical phase shift [12]. Part III is the dead-band compensation. Part II is the PI controller. Here one current sensor is utilized to sense the load current and identify the load resistance.

We compare three control strategies based on the same optimized PI parameters, that is, PI control (termed Control I, using only part II in Figure 7.14), hybrid control (termed Control II, using part I + part II in Figure 7.14), and hybrid control plus dead-band compensation (termed Control III, using part I + part II + part III in Figure 7.14).

In order to investigate the dynamic performance of different control strategies, a simulation is performed where at $t = 0.02$ seconds the output power is changed from 1.2 to $9\,kW$ and at $t = 0.04$ seconds the output power is changed from 9 to $1.2\,kW$, as shown in Figure 7.15. From Figure 7.15c,d, it can be seen that the current modulated by the new proposed algorithm directly enters the steady process with a shorter transient process compared to the traditional PI controller. The PI controller used here is an auxiliary to the dead-band compensator and phase-shift predictor. Therefore the effort on optimizing the PI parameters in much less.

7.4   Minimum Pulse Width (MPW)*

MPW is characterized by the switching speed of power semiconductors. MPW for a MOSFET to switch on/off is on the nanosecond level. It is on the submicrosecond level for an IGBT, and on the microsecond level for an IGCT. Pulses of width less than MPW will lead to insufficient turn-on/off actions of semiconductors, therefore the current and voltage output will fail. So, in practice, MPW is needed to ensure effective switching on/off actions.

However, similar to the dead band, although necessary, the existence of MPW will lead to the distortion of control algorithms. In the voltage source inverter, a number of voltage vectors are distorted and crowded in some narrow regions because of MPW. This is one of the most important factors causing motor mechanical vibrations at low frequency, as shown in Figure 7.16 when the operating frequency is $5\,Hz$ [13].

This is also the reason that most control algorithms, such as vector control and direct torque control, can no longer offer excellent performance when the frequency is low. The distorted vectors will also lead to distorted current and flux linkage, thereby deteriorating the operational performance.

But why is this distortion happening at low-frequency operation? In order to answer this question, we present the regions affected by MPW in one space sector ($60°$ in space) under different switching frequencies, as shown in the darker regions of Figure 7.17. Here t_{MPW} is $50\,\mu s$.

*© [2006] Springer. Reprinted, with permission, from Science in China Series E: Technological Sciences.

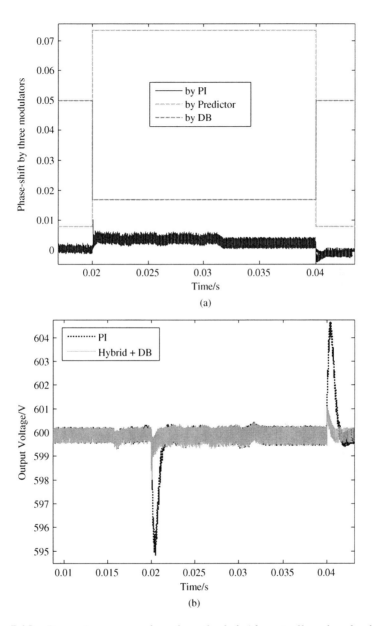

Figure 7.15   Dynamic response based on the hybrid controller plus dead-band compensator: (a) phase shift by three modulators, respectively; (b) voltage variations in the dynamic process; (c) current at t = 0.02 seconds, P = 1.2 kW → 9 kW; and (d) current at t = 0.04 seconds, P = 9 kW → 1.2 kW.

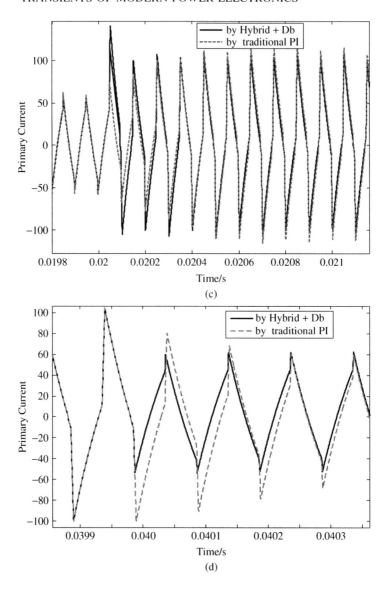

Figure 7.15   (continued)

From Figure 7.17, the conclusions are:

1. The higher the switching frequency, the wider the areas affected by MPW.

2. The larger the MPW, the more severely distorted the vectors.

3. The smaller voltage vectors and edged vectors are much easier to be restricted by MPW.

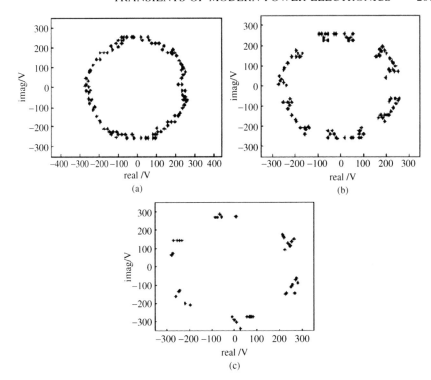

Figure 7.16   The distribution of voltage vectors because of MPW: (a) $t_{MPW} =$ 10 μs, (b) $t_{MPW} = 40$ μs, and (c) $t_{MPW} = 70$ μs.

In the motor control, the flux linkage of the motor should remain constant in most operation modes. Therefore the amplitude of the voltage vector is in direct proportion to the frequency and complies with the following:

$$V \propto \psi \cdot f \tag{7.23}$$

This indicates that, at low frequency, the amplitude of voltage should be small. Therefore small-amplitude vectors will frequently emerge, which will be distorted severely due to MPW. Distorted voltage will cause distorted current, therefore, the system performance will be affected.

MPW not only influences the steady-state operation, but also influences the dynamic processes, for example, the start-up process of a motor drive system.

Example 7.4   MPW in DC Pre-excitation [14, 15]

As we pointed out in Chapter 4, the essence of DC pre-excitation is to inject a DC current into the motor, therefore establishing the flux linkage with some specific value. However, DC pre-excitation is not always effective. Figure 7.18 shows one successful case and one failed case, respectively. In

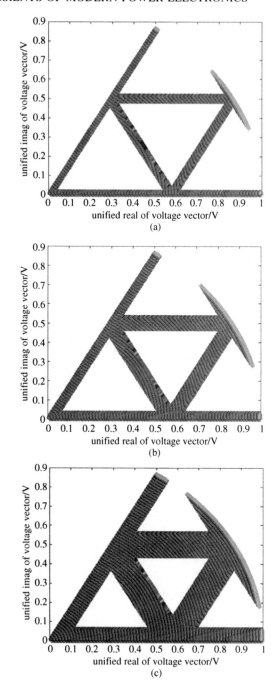

Figure 7.17   Influential regions of MPW: (a) $f_s = 600\,Hz$, (b) $f_s = 1200\,Hz$, and (c) $f_s = 2400\,Hz$.

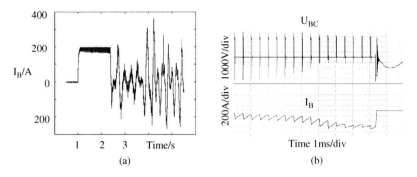

Figure 7.18   Experimental waveforms of the DC pre-excitation: (a) successful case and (b) failed case.

Figure 7.18a, the waveforms are obtained from a 380 V, 160 kW IGBT-based three-level inverter, and the exciting current is maintained at 200 A in the whole pre-exciting process by imposing two voltage vectors, **0NP** and **000**, alternately. In Figure 7.18b, experiments are carried out on the IGCT-based three-level inverter where the pre-exciting current is increasing linearly to exceed the threshold even if the imposed voltage pulses are narrow enough.

The difference lies in MPW. In Figure 7.18a, MPW for the IGBT is 3 μs, while in Figure 7.18b it is 10 μs for the IGCT. Therefore the generated voltage pulse width of Figure 7.18b is always set to 10 μs even if the required pulse width is less than 10 μs. This leads to the actual current larger than the expected current due to MPW.

Therefore, MPW in some cases affects the implementation of control algorithms. This requires the control algorithms to be adaptive to MPW, which is highly associated with the characteristics of the semiconductors. DC pre-excitation is hard to realize for a drive system with large MPW, therefore we could try other control algorithms, for example, AC pre-excitation [14].

7.4.1   Setting the MPW

Compared to the dead-band settings, MPW is complicated. It is highly dependent on the semiconductor, topology, control algorithms, and load condition. For example, in traditional phase-shift control, the duty ratio of the drive signal of each semiconductor is 50%, as shown in Figure 7.9. Suppose the switching frequency is 50 kHz and the turn-on and turn-off pulse widths are equal to 10 μs, which is much larger than the MPW rating of most IGBTs and MOS-FETs as specified by the datasheet. Therefore, for traditional phase-shift control, MPW can be neglected. However, when the switching frequency is higher, the turn-on/off pulse width becomes narrower, therefore approaching the MPW of the semiconductor. Under these circumstances, MPW should be considered.

Here we take the buck circuit shown in Figure 2.5 as an example. Suppose the active switch is an IGCT. For the IGCT 5SHX 14H4502 (4500 V/1100 A), the turn-on and turn-off MPWs are both 10 μs, as mentioned in the relevant datasheet. Also we can get MPW through other key electrical characteristics of the switch. In Figure 7.12, when the semiconductor switch turns on, the interval to fully switch on is $t_{don} + t_r$, defined as the $T_{MPW\ on}$. Similarly, $T_{MPW\ off} = t_{doff} + t_f$. For example, for the MOSFET IXFN 80N50P, $t_{don} = 25$ ns, $t_r = 27$ ns, $t_{doff} = 70$ ns, $t_f = 18$ ns. Therefore, roughly $T_{MPW\ off} = t_{doff} + t_f = 88$ ns and $T_{MPW\ on} = t_{don} + t_r = 52$ ns.

In addition to the semiconductor characteristics (guarantee turning on/off the semiconductors effectively), MPW can also be associated with the dynamic response of the peripheral circuits. It needs to guarantee that at the end of the period, the peripheral circuit has completely reached another steady state. We can introduce another item, Δt, representing the dynamic response time of the peripheral circuit. In Figure 2.5, when DUT turns on, the freewheeling diode turns off. At the end of this turn-on action, both the semiconductor should be turned on fully and the freewheeling diode should end its dynamic process, that is, the reverse recovery. At the same time, this process is greatly affected by the turn-on inductance L_i. Similarly, when DUT turns off, the snubber circuit begins to mitigate the voltage spike inducted by the stray inductance L_{CL}, hence we need to consider the transient process in the snubber circuit besides the current fading process inside the DUT. Therefore the turn-on and turn-off MPWs become

$$T_{MPW\ on} = t_{don} + t_r + \Delta t_1 \tag{7.24}$$

$$T_{MPW\ off} = t_{doff} + t_f + \Delta t_2 \tag{7.25}$$

Here Δt_1 stands for the time for the diode reverse recovery, and Δt_2 represents the time for the dynamic process of the snubber circuit.

Δt_1 and Δt_2 can be obtained by a theoretical calculation if referenced to the datasheet. However, the most precise way to carry out this process is through experiments to probe the periods for reverse recovery of the diode and the snubber circuit. Due to the complex influential factors in the turn-on and turn-off processes, the turn-on and turn-off MPWs are not always the same.

Figure 2.14 shows the turn-off waveforms of one IGCT equipped in the buck converter. From the trailing edge of the control signal to the steady state of the voltage on the IGCT, nearly 30 μs is needed, which is far from the 10 μs mentioned in the datasheet. It only takes 10 μs to turn off the IGCT effectively; however, it needs another 20 μs for the snubber circuit to recover. The whole process can be divided into the following four sections shown in Figure 7.19:

- **Stage 1:** the signal delay. A triggering signal is generated but the IGCT is not activated.

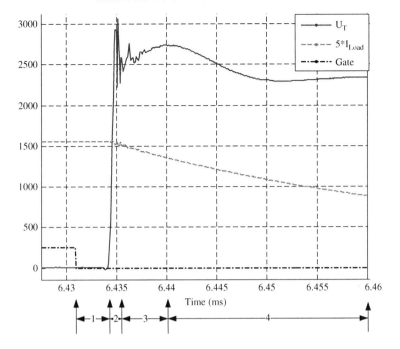

Figure 7.19   The turn-off waveforms of one IGCT.

- **Stage 2:** turn-off stage. The voltage begins to increase when the current drops. Voltage spikes emerge because of the stray inductance.

- **Stage 3:** the voltage of snubber capacitor C_{CL} begins rising because the excessive energy flows from L_{CL} to C_{CL}.

- **Stage 4:** the voltage of snubber capacitor C_{CL} begins dropping when energy stored in C_{CL} begins to transfer to the DC bus through snubber resistor R_s.

7.5   Summary

This chapter mainly focuses on analysis of the dead band and MPW, which are the short-timescale factors determined by the topology and characteristics of semiconductors applied to a real system. From the perspective of pulses, they both belong to the pulse shaping for the gate signals. The difference is that the dead band shapes the time sequence between the two interlocked switches to secure reliable operation of the system, while MPW shapes the width of the gate signal of a single switch to guarantee effective switch on/off actions and restrictions from peripheral circuits. Technically, they both offer reliable operation of the system, the price being distortion of the electrical waveforms and the macroscopic control algorithms. Whether in a DC–AC inverter or in a DC–DC converter, these influences exhibit some similarities.

However, settings of these factors vary with topologies, power rating, semiconductors, applications, and control algorithms. In this chapter we try to address their influences and propose some solutions based on appropriate topologies and control algorithms. It is worthwhile to point out that in some cases their influence is negligible, while others may prohibit effective implementation of control strategies. We consider two examples – the phase-shift control in a DC–DC converter and the DC pre-excitation for a DC–AC inverter – to illustrate their influences and propose relevant solutions to mitigate their side effects.

References

1. Urasaki, N., Senjyu, T., Uezato, K., and Funabashi, T. (2005) An adaptive dead-time compensation strategy for voltage source inverter fed motor drives. *IEEE Transactions on Power Electronics*, **20** (5), 1150–1160.
2. Cardenas, V.M., Horta, S., and Echavarria, R. (1996) Elimination of dead time effects in three phase inverters. IEEE International Power Electronics Congress, pp. 258–262.
3. Liu, H.L. and Cho, G.H. (1993) Three-level space vector PWM in low index modulation region avoiding narrow pulse problem. Power Electronics Specialists Conference, PESC'93, pp. 257–262.
4. Welchko, B.A., Schulz, S.E., and Hiti, S. (2006) Effects and compensation of dead-time and minimum pulse-width limitations in two-level PWM voltage source inverters. Industry Applications Conference, pp. 889–896.
5. Seo, J.H., Choi, C.H., and Hyun, D.S. (2001) A new simplified space-vector PWM method for three-level inverter. *IEEE Transactions on Power Electronics*, **16** (7), 545–550.
6. Li, H., Li, Y., and Ge, Q. (2004) Dead-time compensation of 3-level NPC inverter for medium voltage IGCT drive system. 35th Annual IEEE Power Electronics Specialists Conference, pp. 3524–3528.
7. Aizawa, N., Kikuchi, M., Kubota, H. *et al.* (2010) Dead-time effect and its compensation in common-mode voltage elimination of PWM inverter with auxiliary inverter. Power Electronics Conference (IPEC), pp. 222–227.
8. Bai, H., Zhao, Z., and Mi, C. (2009) Framework and research methodology of short-timescale pulsed power phenomena in high voltage and high power converters. *IEEE Transactions on Industrial Electronics*, **56** (3), 805–816.
9. Mi, C., Bai, H., Wang, C., and Gargies, S. (2008) The operation, design, and control of dual H-bridge based isolated bidirectional DC-DC converter. *IET Power Electronics*, **1** (3), 176–187.
10. Hua, B.A.I., Mi, C.C., and Gargies, S. (2008) The short-time-scale transient processes in high-voltage and high-power isolated bidirectional DC-DC converters. *IEEE Transactions on Power Electronics*, **23** (6), 2648–2656.
11. ABB (2000) Application Note of Integrated Gate Commutated Thyristors.
12. Bai, H., Mi, C., Wang, C., and Gargies, S. (2008) The dynamic model and hybrid phase-shift control of a bidirectional dual active bridge DC-DC converter. IECON'08, pp. 2840–2845.

13. Zheng-ming, Z., Hua, B.A.I., and Liqiang, Y. (2007) Transient of power pulse and its sequence in power electronics. *Science in China Series E: Technological Sciences*, **50** (3), 351–360.

14. Juhasz, G., Halasz, S., and Veszpremi, K. (2000), New aspects of a direct torque controlled induction motor drive. Proceedings of IEEE International Conference on Industrial Technology, Vol. 2, pp. 43–48.

15. Hua, B., Zhengming, Z., Liqiang, Y., and Bing, L. (2006) A high voltage and high power adjustable speed drive system using the integrated LC and step-up transforming filter. *IEEE Transactions on Power Electronics*, **21** (5), 1336–1346.

8

Modulated error in power electronic systems

Modulated error makes real pulses deviate from the intended ones imposed on the objective. The causes and behaviors of modulated errors vary with topology, structure, control algorithm, load condition, and so on. In the real process of carrying out the control algorithm, errors between the real pulse and ideal pulse sometimes will endanger the system. Therefore, strategies to mitigate the modulated error become meaningful, as detailed in this chapter.

8.1 Modulated error between information flow and power flow

In order to facilitate the definition and further understand modulated error in power electronic systems, we use the schematics of Figure 8.1 as an example.

Information flow, for example, the macroscopic control algorithm and its implementation by drive circuits, is the essence of the power electronic system. However, when the information flow passes through a power electronic system, the originally intended power flow can be distorted for a number of reasons:

1. **Complexity of power semiconductors:** the objective of the control algorithm is to control the semiconductor switch, whose nonlinear characteristics may interfere with the control algorithms. In most cases, the information flow needs to be adjusted technically according to the features of semiconductor switches, for example, adding a dead band and setting a minimum pulse width, as mentioned in Chapter 7. In this chapter their influence on the whole system will be further studied.

Transients of Modern Power Electronics, First Edition. Hua Bai and Chris Mi.
© 2011 John Wiley & Sons, Ltd. Published 2011 by John Wiley & Sons, Ltd.

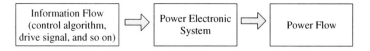

Figure 8.1   Information and power flow in power electronic systems.

2. **Complexity of the power electronic system:** the diversity of topology, structure, and load conditions can affect or distort information flow, which makes the power flow differ from the original intention. A power electronic system is highly nonlinear when it is transitioned from one operational mode to another. For example, the load of adjustable speed drive systems, that is, induction motors, is a highly nonlinear objective.

3. **Complexity of energy flow:** the peripheral circuits and switches coexist in the same system. At any moment the whole system can be in dynamic equilibrium. An energy imbalance can cause extremely high voltage–current transients which may damage the system. Therefore, the ultimate goal is to reach a quasi-static balance of power flow by conducting the excessive accumulated energy in the system in a timely manner. Theoretically, feedback from power flow could be used to adjust the information flow (control algorithm) as shown in Figure 8.2. The differences between the settings and real power flow are defined as modulated error. If the error hits the boundary, the information flow should be adjusted accordingly.

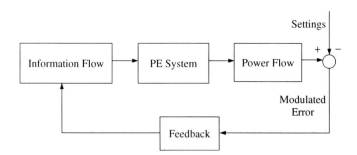

Figure 8.2   Power flow feedback.

There are many types of errors in a practical system. They could be any errors between the reference and actual values, e.g., motor voltage, motor current, motor flux linkage, switch current, switch voltage, and switch turn on/off pulse width. A typical example is when the semiconductor switches are in series connection and the gate signals need to be synchronized, but even the synchronized gate signals may not guarantee that these switches turn on/off simultaneously. The variation of semiconductor characteristics requires additional measures to modify the time sequence of gate signals and additional auxiliary circuit to balance the voltage distribution across series-connected switches. In the following sections, we will discuss the influence of modulated error on the semiconductor itself and the overall system, respectively, and how to correct these errors.

8.2 Modulated error in switching power semiconductors*

The gate signal is responsible for turning on/off the semiconductor switches effectively. However, in practice, the gate signal may not always comply with the power flow. From the rising/trailing edges of a gate signal to the actions of the semiconductors, there is always a turn-on delay or turn-off delay, where the semiconductors remain unchanged although the level of gate signal is altered. This *error* is different for different types of switches and can be fatal in some specific applications and under some special circumstances. One typical example is the series connection of semiconductor switches.

8.2.1 Voltage-balanced circuit for series-connected semiconductors

The input and output voltage of a power electronics converter is limited by the devices' maximum voltage rating. For example, some IGCTs used in induction motor drives can handle a breakdown voltage of about 6 kV. The IGCTs' maximum repetitive voltage is then approximately 5.5 kV, and the output voltage under three-level NPC topology would be 4160 V at the most. Table 8.1 presents the output voltage of a three-level inverter versus the voltage ratings of the semiconductor, reported by ABB Semiconductor.

Table 8.1   The output voltage of a three-level inverter versus voltage ratings of the semiconductor [1].

Output voltage (V)	Half DC-link voltage (V)	Maximum repetitive turn-off voltage (V)
2300	1900	3300
3300	2700	4500
4160	3400	5500
6000	4900	8000
6600	5400	8500
6900	5600	9000
7200	5900	9500

To achieve higher output voltages, it is therefore necessary to connect semi-conductors in series. However, due to potential parameter variation, the voltages across these switches may be different, and some may be damaged due to voltage imbalance. As depicted earlier, synchronous drive signals for series-connected semiconductors produce an asynchronous response due to the different turn-on/off delays. Thus one switch turned off prior to the others in the same series string will inevitably undertake a higher voltage, which causes irreversible damage to the devices due to the unbalanced voltage drops on the series-connected devices. To maintain safe operation of switches in the switching process, equalizing the voltage drops through some extra strategies, either a hardware circuit or software algorithm, is needed.

The methods to handle the voltage imbalance for series-connected semiconductors are summarized as follows:

A. **Voltage balancing circuit** [2, 3]: it is a common practice to use passive circuits, that is, static balancing resistors in the blocking mode and a simple RC snubber circuit in the switching modes [4, 5]. Figure 8.3a illustrates the voltage balancing circuit with static balancing resistors and RC snubber simultaneously.

B. **Active gate design** [6–8]: such a strategy is more appropriate for series-connected IGBTs or MOSFETs, where a voltage sensor is allocated across each device to detect the turn-off voltage (V_{off}). If V_{off} exceeds the reference voltage (V_{ref}), a higher gate voltage will be imposed to reduce V_{off}, as illustrated in Figure 8.3b. The disadvantage of this strategy includes the high cost of the voltage sensors equipped with the switches, and the delay in turning off the switches.

C. **Gate clamping circuit** [9, 10]: Figure 8.3c shows a semiconductor gate clamping circuit. If the voltage drop exceeds the threshold, a Zener diode will break down with current flowing into the gate to charge the gate. This consequently decreases the voltage drop.

D. **Combination of A and B** [9]: the design procedures for method A (*voltage balancing circuit*) reported in the previous literature are relatively simple and widely used in many applications. The RC snubber remains effective as well as low cost. In this process, several key issues need to be addressed:

1. The precise model of semiconductor switches, both controllable switches and diodes. This is needed in order to estimate the influence of the modulated error caused by the variation of semiconductor switches and optimize the RC values. Ideal models of the semiconductors cannot reveal the variations of switch characteristics and therefore are not helpful in eliminating the voltage imbalance. Ideally, this model needs to reveal the difference in turn-on/off delay, approach the real switching process of semiconductors, and behave precisely under different load conditions.

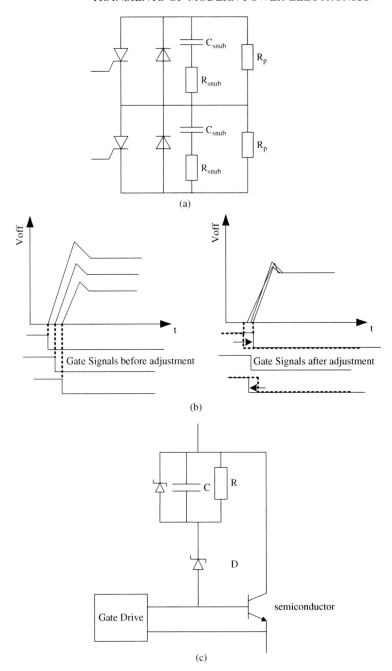

Figure 8.3   The methods to deal with voltage imbalance in series connected switches: (a) the balancing circuit, (b) the active gate design for Method B, and (c) the semiconductor gate clamping circuit.

2. Impact of the RC circuit on the switches. Inclusion of the peripheral circuit will impact the switches in the switching-on/off processes and change the characteristics of the switches significantly. For example, the additional di/dt impact by C in the switching-on process requires the snubber resistor R to be sufficiently large. However, a high resistor value will result in the system incapable of balancing the voltage across different switches. In other words, adding this circuit decreases the voltage imbalance in the turning-off process, but should not make these impacts exceed the electrical endurance of the semiconductors.

3. Loss is a side effect of RC circuit design. Theoretical analysis shows that the loss due to a dynamic balancing circuit in one switching period is

$$Loss_{snubber} = C_d V^2 \qquad (8.1)$$

where C_d is the snubber capacitance and V is the off-state voltage across one IGCT module. The RC snubber will decrease the turn-off loss of the switches, but will result in additional loss in the circuit itself. When an IGCT turns on, the energy stored in the capacitor is drained out. When an IGCT turns off, this capacitor is recharged. Consider two series-connected IGCTs working under 6000 V/800 A as an example. The simulated loss distribution based on the functional model of an IGCT is shown in Figure 8.4.

4. The control algorithm must guarantee that the on/off pulse widths are large enough to fully turn on/off the IGCTs. This is also the minimum pulse width required.

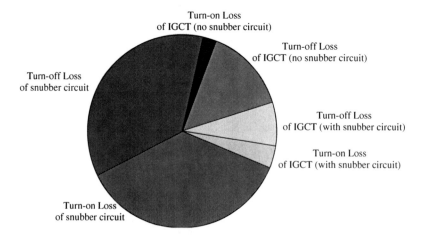

Figure 8.4   Loss comparison with/without RC snubber circuit.

The parameter selection of a RC snubber is illustrated in Figure 8.5 and Table 8.2.

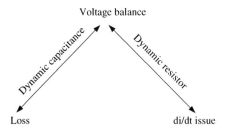

Voltage balance

Dynamic capacitance

Dynamic resistor

Loss   di/dt issue

Figure 8.5   Tradeoffs in the design. In Table 8.2, ↑, ↓, and − mean increasing, decreasing, and no influence, respectively.

Table 8.2   Influential factors in this design [11].

	di/dt	I_{max}	ΔV	Loss
C	−	↑	↓	↑
R	↓	↓	↑	−

The experimental results are shown in Figure 8.6. When no snubber circuit is equipped, the voltage distribution varies drastically as shown in Figure 8.6a,c. $C = 1\,\mu F$ and $R = 1\,\Omega$ lead to a smaller voltage difference in both the series-connected IGCTs and the diodes as shown in Figure 8.6b,d. To see how the optimal solution is obtained, readers can refer to [11] for more details.

8.2.2   Accompanied short-timescale transients

The RC snubber aims to readjust the modulated error by equalizing the voltage on different switches and eventually reaches a high output voltage. However, adding the RC snubber circuit changes the circuitry topology of the overall system, which not only increases loss, but also contributes to other accompanying short-timescale phenomena in the original commutating process.

8.2.2.1   Turn-off voltage waveforms of a single switch

The turn-off voltage waveform across a single IGCT with a RC snubber is shown in Figure 8.7. In this waveform, there are four main steps:

- **Step 1:** a negative voltage across the IGCT emerges with 20 V amplitude, equal to the gate drive voltage. Figure 8.8 shows the IGCT structure in the turn-off process, where −20 V is imposed on junction J3 in the turn-off process to commutate the current from the IGCT to the gate. In this stage the IGCT is still on, and the junctions J1 and J2 do not withstand

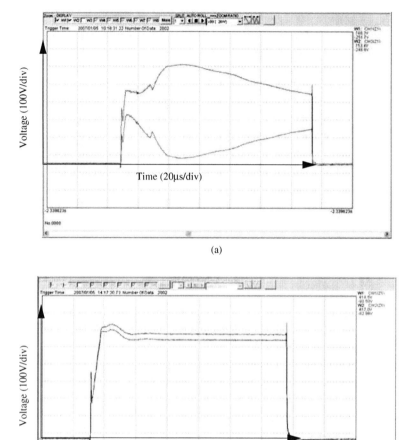

(a)

(b)

Figure 8.6   Comparison of dynamic voltage equilibrium under different RC parameters: (a) IGCTs, no RC circuit; (b) IGCTs, with RC circuit ($R_d = 1\Omega$, $C_d = 1\,\mu F$); (c) diodes, no RC circuit; and (d) diodes, with RC circuit ($R_d = 1\Omega, C_d = 1\,\mu F$).

any voltage. Therefore the voltage across the anode and cathode is $-20\,V$, which makes current flow reversely in the dynamic voltage balancing circuit. This voltage is in fact conducted through the gate-drive circuit and does no harm to the system.

- **Step 2:** both i and di/dt through the IGCT increase, which induces the first voltage spike via the stray inductance of the linking wires or bus bar.

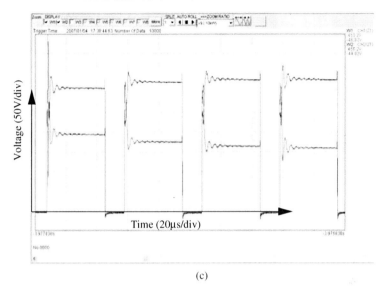

(c)

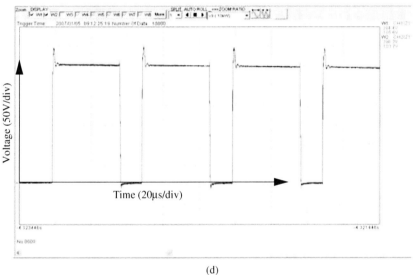

(d)

Figure 8.6   (continued)

Simulation shows that this voltage spike is closely related to the stray inductance located inside the RC snubber. In real applications, an inductance-free
resistor and capacitor are required.

- **Step 3:** current i increases while di/dt decreases. The voltage across the
  IGCT is mainly the voltage drop on the resistor and capacitor.

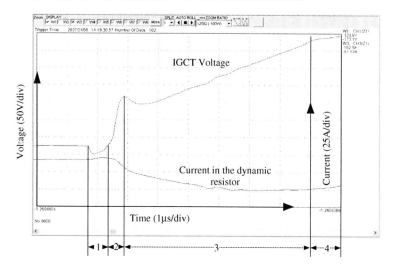

Figure 8.7   The turn-off voltage and current waveforms.

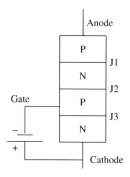

Figure 8.8   The IGCT structure.

- **Step 4:** current i decreases. However, the capacitor is continuously charged. Therefore the voltages in steps 3 and 4 are both increasing until they reach steady state.

Overall, a RC snubber mitigates the turn-off voltage issue of the semiconductors. If R is chosen appropriately, the impact of current during turn-on can also be reduced.

8.2.2.2 Current discontinuity in the dead band

Consider the three-level topology shown in Figure 8.9, and assume that "1" stands for the on state and "0" for the off state of an IGCT. For example, 1100 stands for Sa1 and Sa2 on and Sa3 and Sa4 off, where each Sai, $i = (1-4)$, represents the two IGCTs in series connection. The load current is $I_L > 0$ (from the bridge to the load). Liu *et al.* [12] reported that the current in Sa2 is intended to decline in the dead band when the topology changes from [1100] to [0110], that is, the intermediate state [0100]. When $I_L > 0$, the current decline in Sa2 is $I_L/4$. The topology variation and voltage distribution of states [1100] and [0100] are shown in Figure 8.9a and b, respectively. Figure 8.9c shows the current flow and voltage change in the transient [1100] → [0100] through the simplified circuitry topology. The initial and final states of the capacitor voltage are indicated respectively.

The voltages across the capacitors are

$$V_1 = \frac{1}{C} \int_0^t i_1 \, dt$$

$$V_3 = \frac{1}{2} V_{DC} - \frac{1}{C} \int_0^t i_3 \, dt$$

$$V_4 = \frac{1}{2} V_{DC} - \frac{1}{C} \int_0^t i_5 \, dt \qquad (8.2)$$

$$V_{D_1} = \frac{1}{2} V_{DC} - \frac{1}{C} \int_0^t i_2 \, dt$$

$$V_{D_2} = \frac{1}{C} \int_0^t i_4 \, dt$$

Based on Kirchhoff's current law, Figure 8.9c can be mathematically formulated as

$$i_1 + i_2 + i_3 = I_L$$
$$i_4 + i_5 = i_3 \qquad (8.3)$$

Based on Kirchhoff's voltage law, the voltage equation for loop 1 is

$$\frac{1}{2} V_{DC} - i_1 R - V_1 + i_2 R - V_2 = 0 \qquad (8.4)$$

For loop 2

$$\frac{1}{2} V_{DC} - i_4 R - V_{D2} + i_5 R - V_4 = 0 \qquad (8.5)$$

For loop 3

$$V_{DC} - i_1 R - V_1 + i_3 R + i_5 R - V_3 - V_4 = 0 \qquad (8.6)$$

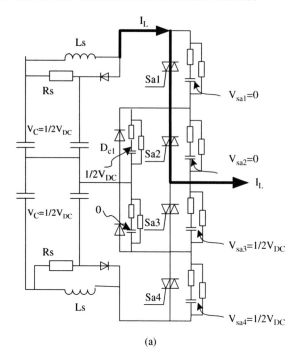

(a)

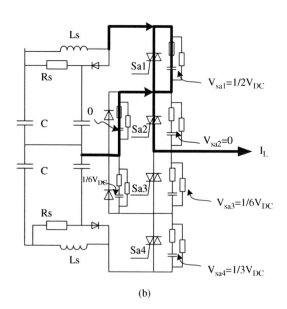

(b)

Figure 8.9   Current changes in 1100 → 0100 under different conditions: (a) current at [1100], (b) current in [1100]→[0100], and (c) simplified topology for [1100]→[0100].

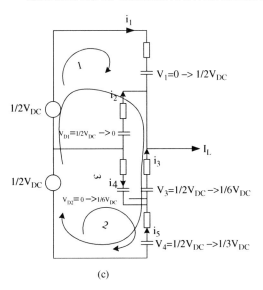

(c)

Figure 8.9 (continued)

Suppose that R_d is small, DC-bus voltage (V_{DC}) is high, and load current (I_L) is small. Then the voltage drop on the snubber resistors can be neglected. Consequently the capacitors will be charged or discharged with constant current. Solutions to Equations 8.2–8.6 are

$$i_1 = \frac{3}{8}I_L \quad i_2 = \frac{3}{8}I_L \quad i_3 = \frac{1}{4}I_L \quad i_4 = \frac{1}{8}I_L \quad i_5 = \frac{1}{8}I_L \tag{8.7}$$

Accordingly the current in Sa2 is $i_1 + i_2 = \frac{3}{4}I_L$ in the dead band [0100]. This value is different from I_L when RC snubber is not included. The difference is $\frac{1}{4}I_L$.

However, if

$$\frac{V_{DC}}{2} - i_2 \times 2R_d < 0 \tag{8.8}$$

that is, $R_d > \frac{2}{3}V_{DC}/I_L$ referring to Equation 8.7, the voltage drop on the snubber resistors cannot be neglected. The absolute value of current decline in Sa2 would be less than $I_L/4$ and the constant current discharging process will disappear. In this case, the current declination can be calculated by the following equation (see [9] for a detailed derivation):

$$\Delta i_{Sa2/3} = \begin{cases} -\dfrac{1}{4}I_L & R_d \le \dfrac{2}{3}\dfrac{V_{DC}}{I_L} \\[3mm] -\dfrac{V_{DC}}{6R_d} & R_d > \dfrac{2}{3}\dfrac{V_{DC}}{I_L} \end{cases} \tag{8.9}$$

When $R_d > \frac{2}{3}V_{DC}/I_L$, the increase of R_d would be followed by a decrease of current declination. Figure 8.10a,b show the current decline in Sa2 with different R_d and C_d, which validates Equation 8.9. Simulation also shows that this current drop is not very sensitive to the value of snubber capacitance C_d.

8.2.2.3 Analysis of the RC circuit under different conditions

Assume all the following processes are in the three-level bridge shown in Figure 8.9.

Since the commutating process in a DC–AC inverter for motor drives is highly dependent on the load current direction, in the following analysis the operation modes will be classified according to the current directions:

1. *For $I_L > 0$:*
   Case 1A, 0100 → 0110. When Sa3 is turned on, the current in Sa3 is given by

   $$i_{Sa3}(t) = \frac{V_{DC}}{3R_d}e^{-t/(R_dC_d)} + \frac{2V_{DC}}{3\sqrt{R_d^2 - 4\frac{L_s+L_i}{C_d}}}$$

   $$\times \left\{ e^{-\frac{-R_d+\sqrt{R^2-[4(L_s+L_i)/C_d]}}{2L_s}t} - e^{-\frac{-R_d-\sqrt{R^2-[4(L_s+L_i)/C_d]}}{2L_s}t} \right\}$$

   $$(8.10)$$

   The first and second terms of this current are mainly attributed to the snubber capacitance and inductance, respectively. The topology of this process is shown in Figure 8.11.
   Case 1B, 0100 → 1100. In this case, the current impact is similar to that described by Equation 8.10. The current impact on Sa1 is the most significant but smaller than that in *Case 1A*. The topology is shown in Figure 8.12. Similarly, the current expression of cases 0011 → 0010 and 0010 → 0011 can be derived.

2. *For $I_L < 0$:*
   Case 2A, 0110 → 0100. Neglect the influence of L_s and assume that D_1 (the anti-parallel diode of IGCT Sa1) is blocked in the commutating process, as shown in Figure 8.13.
   The current in the RC circuit of Sa1 is $-I_L/3$ while that in Sa3 is $I_L/3$. The condition of this phenomenon is similar to Equation 8.8. That is, if $R_d < (3V_{DC}/2I_L)$, C_d would be discharged with constant current $I_L/3$. This period lasts for $\Delta t = 3C_d[(V_{DC}/2) - (I_LR_d/3)]/I_L$. After this period Δt, the RC snubber of Sa1 is shorted and the anti-parallel diode of Sa1 conducts. The snubber circuit of Sa3 will be charged from zero to $V_{DC}/2$

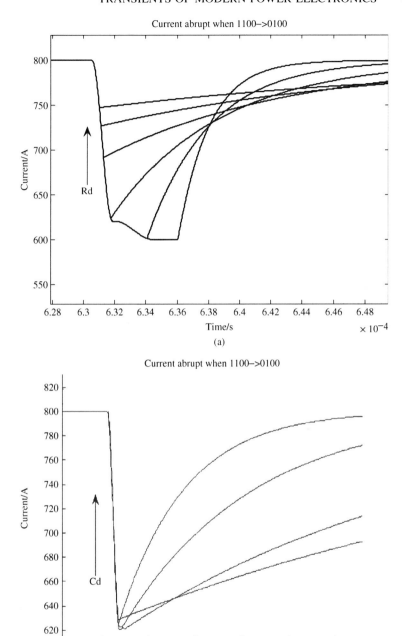

Figure 8.10   Current in Sa2, 1100 → 0100 at $I_L > 0$: (a) with different R_d and (b) with different C_d.

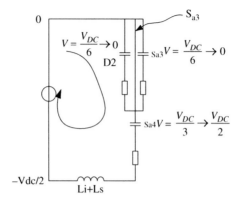

Figure 8.11   The topology of 0100 → 0110 at $I_L > 0$.

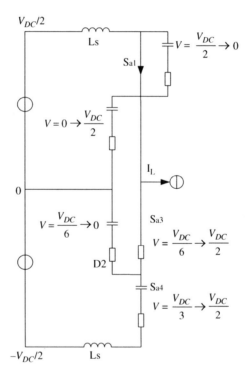

Figure 8.12   The topology of 0100 → 1100 at $I_L > 0$.

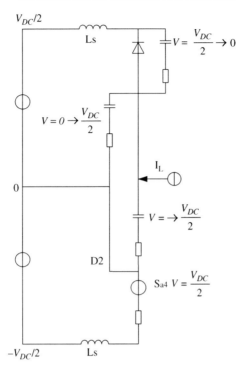

Figure 8.13   The topology of 0110 → 0100 at $I_L < 0$.

with constant current $I_L/3$. On the other hand, if $R_d > (3V_{DC}/2I_L)$, $\Delta t = 0$; that is, no constant current-charging process emerges [13].

As analyzed in Figure 8.9, when $I_L < 0$, current declination happens to Sa1 when 1100 → 0100. Figure 8.14 shows its counterpart when a similar transient happens to the freewheeling diode of Sa1 when 0110 → 0100 at $I_L < 0$. When $R_d > 7.9\Omega\,(V_{DC} = 9400\text{ V}, I = 800\text{ A})$ the process of constant current discharge does not exist, which is consistent with the previous analysis.

Case 2B, 0010 → 0110 and 0010 → 0011. These are two other conditions where a large current will appear in Sa2 and Sa4, respectively. The methodology of investigation is the same as above.

8.3   Modulated error in the DC–AC inverter

Section 8.2 analyzed the compensation of modulated error due to the semiconductor itself, which is at the device level. When we extend this to the system level, the impact of modulated error on the system is worth studying. For example,

Current abrupt in Sa1 snubber when 0110□ >0100

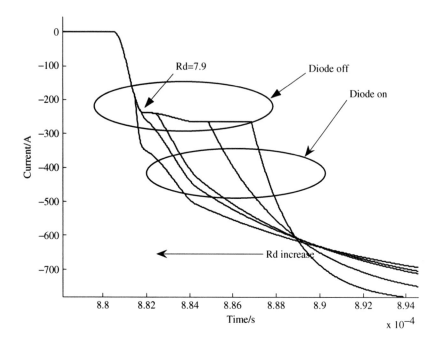

Figure 8.14   Current in diode of Sa1, 0110 → 0100 at $I_L < 0$.

modulated error caused by the minimum pulse width in low-speed operation, that is, distortion of the voltage vectors, has been addressed in Chapter 7. Not only does modulated error exist in some specific regions, but it also occurs throughout the whole operational range. It has been observed that when the 6000 V, 1250 kW NPC inverter shown in Figure 3.9 is operated at 50 Hz, there is a current oscillation with a period of 30 minutes, as shown in Figure 8.15a [14].

This abnormal phenomenon is attributed to the minimum pulse width. When the control algorithm is implemented by a microcontroller, the switching frequency is 600 Hz and the fundamental frequency is 50 Hz. Hence the whole space is divided into 600/50 = 12 parts. Ideally the calculated angle of each space voltage vector in every switching period is fixed. For example, the angle of the first space vector is 30°, so the second one is 30° + 360°/12 = 60°. If the microcontroller could implement the code without error, the angle of each space vector in every switching period would be fixed.

However, calculation errors can cause inaccuracy in the angle calculations. Specifically, each sector is mathematically 360°/12 = 30°. But in a microcontroller the angle may not be precisely 30° due to resolution and limited number of bits of each number; that is, it could be 30.01°. Then the reference voltage vector might be located at 30° in the first 0.02 seconds (1/50 Hz). After another 0.02 seconds, the angle accumulates to 30° + 30.01° × 12 − 360° = 30.12°, to

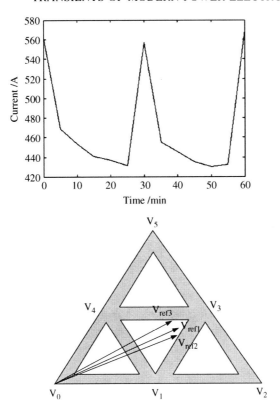

Figure 8.15   Modulated error caused by the minimum pulse width: (a) current oscillation and (b) variation of the voltage space vectors.

$30.24°$ for the third 0.02 seconds, and so on. Therefore, after some time, the voltage vector will shift to the regions restrained by the minimum pulse width, shown as V_{ref2} or V_{ref3} in Figure 8.15b, which is not the same as V_{ref1}. As mentioned in Chapter 7, when the voltage vector slides into the restrained region of minimum pulse width, the voltage waveform will be distorted which leads to distorted current when combined with the LC filter shown in Figure 3.9.

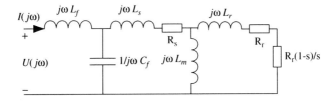

Figure 8.16   Equivalent circuit of the LC filter, transformer, and motor.

The equivalent circuit of the LC filter, the motor and the primary side of the isolation transformer is shown in Figure 8.16. L_f is the leakage inductance of the transformer from the primary side to the secondary side, C_f is the smoothing capacitance, L_s is the stator leakage inductance of the induction motor, R_s is the stator resistance, L_m is the equivalent excitation inductance of the motor, L_r is the rotor leakage inductance, R_r is the rotor resistance, and s is the slip ratio. $U(j\omega)$ and $I(j\omega)$ are the phase voltage and current of the inverter in the frequency domain, respectively.

The impedance of this equivalent circuit is

$$Z(j\omega) = \frac{U(j\omega)}{I(j\omega)} = j\omega L_f + \frac{1}{j\omega C_f}||Z_{motor}(j\omega)$$

$$Z_{motor}(j\omega) = j\omega L_s + R_s + j\omega L_m|| \left(j\omega L_r + \frac{R_r}{s} \right)$$

(8.11)

From Equation 8.11, the network impedance can be derived and plotted as shown in Figure 8.17a, where the valley point of this impedance curve emerges around 300 Hz in the frequency domain. If there is a harmonic voltage around 300 Hz, the current of relevant harmonics will be amplified by the low impedance, as shown in the comparison of Figure 8.17b,c.

Therefore the oscillation in Figure 8.15a is attributed to the combination of three factors: the minimum pulse width, modulated error, and peripheral LC filter. If there is no LC filter, the characteristics of the motor impedance will not allow this oscillation to happen. If the space vectors are not calculated on-line, but allocated with fixed angles stored in a table, they will never enter the region restrained by the minimum pulse width, therefore the current distribution will only behave as shown in Figure 8.17b, not Figure 8.17c. At this point, selected harmonics elimination PWM (SHEPWM) has been validated as the feasible solution for this type of modulated error [15].

8.4   Modulated error in the DC–DC converter*

Modulated error also exists in DC–DC converters. The principles applied to analyze the modulated error in the DC–AC inverter can be applied to a DC–DC converter, although the in-depth causes of modulated error can vary. In Section 7.4, when voltage vectors hit the boundary lines of the regions restrained by the minimum pulse width, the overall system can be regarded as working under the boundary conditions. In a DAB-based converter with phase-shift control as depicted in Chapter 7, the minimum pulse width is not the primary concern. However, microscopic and macroscopic factors still coexist in this system. It has its own boundary operation modes.

* © [2008] IEEE. Reprinted, with permission, from IECON'08.

As shown in the previous chapter, the peak current of the DAB-based DC–DC converter when $V_2 \geq n V_1$ can be written as

$$i_{peak} = \frac{n}{2\omega L_s}[-(\pi - 2\phi)n V_1 + V_2\pi] = \frac{n}{4 f_s L_s}[-n V_1 + 2Dn V_1 + V_2] > 0 \tag{8.12}$$

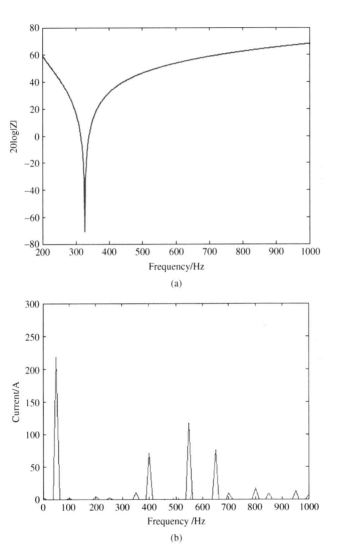

(a)

(b)

Figure 8.17   Influence of minimum pulse width and the LC characteristics: (a) impedance characteristics of the LC filter, (b) current spectrum (no minimum pulse width), and (c) current spectrum (with minimum pulse width).

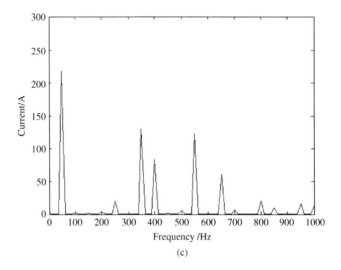

Figure 8.17 (continued)

Solving D in Equation 7.20 and substituting it into Equation 8.12, the peak current to be found:

$$i_{peak} = \frac{n}{4f_s L_s}\left[-nV_1 + \left(1 - \sqrt{1 - \frac{8f_s L_s P}{nV_1 V_2}}\right)nV_1 + V_2\right]$$

$$= \frac{n}{4f_s L_s}\left[V_2 - \sqrt{1 - \frac{8f_s L_s P}{nV_1 V_2}}\,nV_1\right]$$

(8.13)

In practical applications, the system is expected to undertake the smallest peak current at the rated power. In order to get the global minimum value of i_{peak}, let

$$\frac{\partial i_{peak}}{\partial L_s} = -\frac{nV_2}{4f_s L_s^2} + \frac{n^2 V_1}{4f_s L_s^2}\frac{1}{2}\frac{2 - \dfrac{8f_s L_s P}{nV_1 V_2}}{\sqrt{1 - \dfrac{8f_s L_s P}{nV_1 V_2}}} = -\frac{nV_2}{4f_s L_s^2} + \frac{n^2 V_1}{4f_s L_s^2}\frac{1}{2}\frac{1+\Delta}{\sqrt{\Delta}} = 0$$

(8.14)

Here

$$\Delta = 1 - \frac{8f_s L_s P}{nV_1 V_2} > 0$$

Consequently, when

$$\frac{1}{2}\frac{1+\Delta}{\sqrt{\Delta}} = \frac{V_2}{nV_1}$$

the peak current of the primary side will reach a minimum at rated power P.

Considering the condition for $V_2 < nV_1$, the peak current for the entire operating range can be found. It will reach the valley point when

$$L_s = \frac{nV_1V_2}{4fsP}(1 - m^2 + m\sqrt{m^2 - 1}) \quad (m = \max(V_2, nV_1)/\min(V_2, nV_1)) \tag{8.15}$$

The minimum of the primary current peak is

$$i_{peak\,\min} = \frac{nP}{\max(nV_1, V_2)(1 - m^2 + m\sqrt{m^2 - 1})} \tag{8.16}$$
$$\left(m - \sqrt{2m^2 - 2m\sqrt{m^2 - 1} - 1} - 1\right)$$

The relationships between $i_{\max}$ in Equation 8.16 and leakage inductance L_s are shown in Figure 8.18.

Compared to Figure 8.18a,c, $V_2 = nV_1$ is a good choice in the steady state operation where a smaller leakage inductance is required, as shown in Figure 8.18e. When $L_s = 15\,\mu H$, $nV_1 = V_2 = 600$ V, the peak current on the primary side is only 33 A. However, the output voltage V_2 is not exactly 600 V, which could be caused by the control algorithm, dead band, or dynamic load disturbance. If there is a 10 V error, that is, V_2 changes from 600 to 590 or 610 V, the primary current will jump from 33 to 66–67 A, as shown in Figure 8.18a–d. This amplitude has exceeded the current thresholds determined by the IGBTs. In the dynamic process, this current peak is expected to be even higher. Experimental waveforms are shown in Figure 8.18g for the current oscillation when $nV_1 = V_2$.

From Equation 8.12, the maximum inductor current is [16]

$$i_{peak} = \frac{nV_1}{4f_sL_S}[2D + |m - 1|] \tag{8.17}$$

Assuming there is a disturbance ΔD on the phase-shift duty ratio D, the variation of maximum current is

$$\Delta i_{peak} = \frac{i_{peak}(D + \Delta D) - i_{peak}(D)}{i_{peak}(D)} = \frac{2\Delta D}{2D + |m - 1|} \tag{8.18}$$

Here $\Delta D = D^* - D$, where D^* is the reference phase-shift duty and D is the real value. When $m = 1$

$$\Delta i_{peak}\% = \frac{\Delta D}{D} \tag{8.19}$$

Equation 8.19 indicates that the current oscillation on the boundary condition is actually in direct proportion to the modulated error by the microcontroller. Therefore the boundary mode where $m = V_2/nV_1 = 1$ is an unstable operating point.

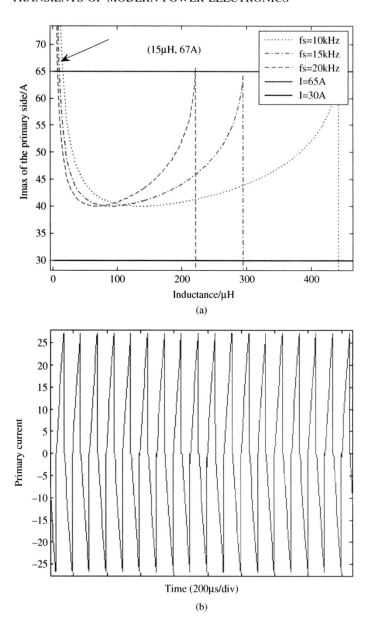

Figure 8.18   Peak current vs. leakage inductance under different parameter choices: (a) $nV_1 > V_2$*; (b)* $nV_1 > V_2$*, simulation; (c)* $nV_1 < V_2$*; (d)* $L_s = 15\,\mu H$*,* $nV_1 < V_2$*, simulation; (e)* $nV_1 = V_2$*; (f)* $L_s = 15\,\mu H$*,* $nV_1 = V_2$*, simulation; (g)* $nV_1 = V_2$*; and (h)* $nV_1 > V_2$*, simulated waveforms.*

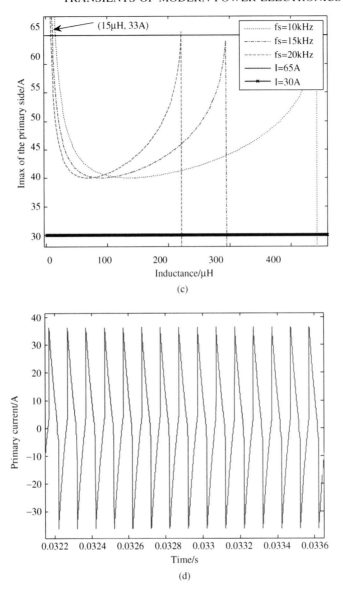

Figure 8.18   (continued)

To solve the conflicts caused by the boundary mode, we need a two-step solution. First, redesign the isolated transformer and make $nV_1 \neq V_2$. For example, use $n = 1.9$ or $n = 2.1$ instead of $n = 2$ for the case of $V_1 = 300$ V and $V_2 = 600$ V. Second, enlarge the leakage inductance to improve the system dynamic performance, but it will result in extra loss and decrease system efficiency.

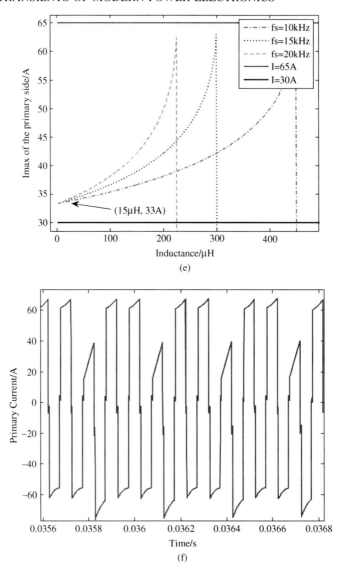

(e)

(f)

Figure 8.18   (continued)

In most cases, quantitative analysis is needed in order to eliminate modulated error. For this purpose, a mathematical model needs to be established. Consider the simplified DAB converter as an example (Figure 8.19). To simplify the analysis, all the switches are assumed to be ideal. The transformer is replaced with its equivalent leakage inductance L and n is the turns ratio of the transformer. Voltages on the two sides of the transformer are nv_1 and v_2, respectively; v_s is the voltage of the power supply and r_s is the inner resistance of the primary power supply.

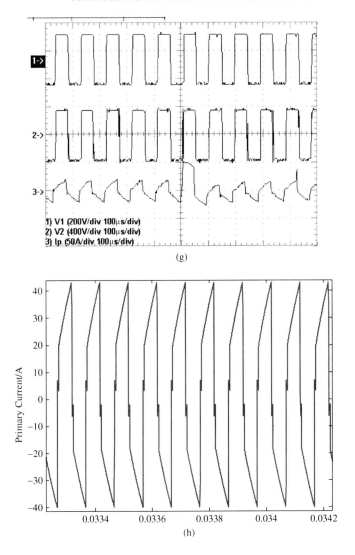

1) V1 (200V/div 100μs/div)
2) V2 (400V/div 100μs/div)
3) Ip (50A/div 100μs/div)

(g)

(h)

Figure 8.18   (continued)

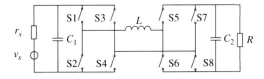

Figure 8.19   Simplified circuit of a DAB DC–DC converter.

There are four sets of equations in one switching period. In each sub-interval shown in Figure 7.5, the equation has a linear form of $\dot{x}(t) = A_i x(t) + B_i u(t)$:

$$
\begin{bmatrix} \dfrac{di_L}{dt} \\[2mm] \dfrac{dv_1}{dt} \\[2mm] \dfrac{dv_2}{dt} \end{bmatrix} = \begin{bmatrix} 0 & \dfrac{1}{L} & \dfrac{1}{L} \\[2mm] -\dfrac{1}{C_1} & -\dfrac{1}{r_s C_1} & 0 \\[2mm] -\dfrac{1}{C_2} & 0 & -\dfrac{1}{RC_2} \end{bmatrix}
$$

$$
\begin{bmatrix} i_L \\ v_1 \\ v_2 \end{bmatrix} + \begin{bmatrix} 0 \\[2mm] \dfrac{1}{r_s C_1} \\[2mm] 0 \end{bmatrix} v_s \quad t \in [\, 0 \quad DT_s \,] \tag{8.20}
$$

$$
\begin{bmatrix} \dfrac{di_L}{dt} \\[2mm] \dfrac{dv_1}{dt} \\[2mm] \dfrac{dv_2}{dt} \end{bmatrix} = \begin{bmatrix} 0 & \dfrac{1}{L} & -\dfrac{1}{L} \\[2mm] -\dfrac{1}{C_1} & -\dfrac{1}{r_s C_1} & 0 \\[2mm] \dfrac{1}{C_2} & 0 & -\dfrac{1}{RC_2} \end{bmatrix}
$$

$$
\begin{bmatrix} i_L \\ v_1 \\ v_2 \end{bmatrix} + \begin{bmatrix} 0 \\[2mm] \dfrac{1}{r_s C_1} \\[2mm] 0 \end{bmatrix} v_s \quad t \in [\, DT_s \quad T_s \,] \tag{8.21}
$$

$$
\begin{bmatrix} \dfrac{di_L}{dt} \\[2mm] \dfrac{dv_1}{dt} \\[2mm] \dfrac{dv_2}{dt} \end{bmatrix} = \begin{bmatrix} 0 & -\dfrac{1}{L} & -\dfrac{1}{L} \\[2mm] \dfrac{1}{C_1} & -\dfrac{1}{r_s C_1} & 0 \\[2mm] \dfrac{1}{C_2} & 0 & -\dfrac{1}{RC_2} \end{bmatrix}
$$

$$
\begin{bmatrix} i_L \\ v_1 \\ v_2 \end{bmatrix} + \begin{bmatrix} 0 \\[2mm] \dfrac{1}{r_s C_1} \\[2mm] 0 \end{bmatrix} v_s \quad t \in [\, T_s \quad (1+D)T_s \,] \tag{8.22}
$$

$$
\begin{bmatrix} \dfrac{di_L}{dt} \\[2mm] \dfrac{dv_1}{dt} \\[2mm] \dfrac{dv_2}{dt} \end{bmatrix} = \begin{bmatrix} 0 & -\dfrac{1}{L} & \dfrac{1}{L} \\[2mm] \dfrac{1}{C_1} & -\dfrac{1}{C_1 r_s} & 0 \\[2mm] -\dfrac{1}{C_2} & 0 & -\dfrac{1}{RC_2} \end{bmatrix}
$$

$$
\begin{bmatrix} i_L \\ v_1 \\ v_2 \end{bmatrix} + \begin{bmatrix} 0 \\[2mm] \dfrac{1}{C_1 r_s} \\[2mm] 0 \end{bmatrix} v_s \quad t \in [\, (1+D)T_s \quad 2T_s \,] \tag{8.23}
$$

The state variable i_L varies faster than v_1 and v_2. So the system belongs to a two-timescale system. If we only concern about the voltage response over the switching period other than the dynamic behaviors with the smaller timescale, it is necessary to develop an averaged model for the slow state variables, that is, v_1 and v_2. The averaged model is a reduced-order model and is simpler compared to the full order model in Equations 8.20–8.23, which belongs to variable structure systems (VSRs) from the control theory point of view.

To derive an averaged model for the slow state variables, we first assume that the slow variables v_1 and v_2 behave more like constants compared to the fast variable (i_L). Second, we divide the system into two subsystems, one for the slow-variable system and the other for the fast-variable system:

$$\frac{dX}{dt} = A_{sx}^{(i)} X + A_{sy}^{(i)} Y + B_s^{(i)} U$$

$$\frac{dY}{dt} = A_{fx}^{(i)} X + A_{fy}^{(i)} Y + B_f^{(i)} U$$

$$(8.24)$$

where $i = 1$–4.

Third, the fast variables in the above equations of the slow system should be replaced according to Equations 8.20–8.23 by substituting i_L into the slow variables v_1 and v_2 in the corresponding mode, as shown in Equations 8.25 and 8.26 below. Due to the symmetry of the waveforms, only expressions in $[0, T_s]$ are listed. Note that $T_s = 1/2f_s$.

$$C_1 \frac{dv_1}{dt} = \begin{cases} \frac{v_s}{r_s} - \frac{v_1}{r_s} - \frac{v_1+v_2}{L}t - \frac{1}{4f_sL}[(1-2D)v_2 - v_1] & [0, DT_s] \\ \frac{v_s}{r_s} - \frac{v_1}{r_s} - \frac{v_1-v_2}{L}(t - dT_s) - \frac{1}{4f_sL}[v_2 - (1-2D)v_1] & [DT_s, T_s] \end{cases}$$

$$(8.25)$$

$$C_2 \frac{dv_1}{dt} = \begin{cases} -\frac{v_1+v_2}{L}t - \frac{1}{4f_sL}[(1-2D)v_2 - v_1] - \frac{v_2}{R} & [0, DT_s] \\ \frac{v_1-v_2}{L}(t - dT_s) + \frac{1}{4f_sL}[v_2 - (1-2D)v_1] - \frac{v_2}{R} & [DT_s, T_s] \end{cases}$$

$$(8.26)$$

Averaging v_1 and v_2 in $[0, T_s]$, and rewriting the formula,

$$\begin{bmatrix} \frac{d\langle v_1 \rangle}{dt} \\ \frac{d\langle v_2 \rangle}{dt} \end{bmatrix} = \begin{bmatrix} -\frac{1}{r_sC_1} & \frac{(D^2 - D)}{2Lf_sC_1} \\ \frac{(-D^2 + D)}{2Lf_sC_2} & -\frac{1}{RC_2} \end{bmatrix} \begin{bmatrix} \langle v_1 \rangle \\ \langle v_2 \rangle \end{bmatrix} + \begin{bmatrix} \frac{1}{r_sC_1} \\ 0 \end{bmatrix} \langle v_s \rangle$$

$$(8.27)$$

From Equation 8.27, the matrix is negatively defined. Hence the whole system is stable and reaches final steady state when and only when

$$\frac{d\langle v_1\rangle}{dt} = \frac{d\langle v_2\rangle}{dt} = 0$$

that is,

$$v_1 = \frac{v_s}{1 + Rr_s\left(\frac{D(1-D)}{2Lf_s}\right)^2} \quad \text{and} \quad v_2 = \frac{v_s}{1 + Rr_s\left(\frac{D(1-D)}{2Lf_s}\right)^2}\frac{RD(1-D)}{2Lf_s} \quad (8.28)$$

Now, assuming there is a small variation of duty ratio Δd, the small-signal model is

$$\begin{bmatrix} \dfrac{d\langle v_1\rangle}{dt} \\ \dfrac{d\langle v_2\rangle}{dt} \end{bmatrix} = \begin{bmatrix} -\dfrac{1}{r_sC_1} & \dfrac{(D^2-D)}{2Lf_sC_1} \\ \dfrac{(-D^2+D)}{2Lf_sC_2} & -\dfrac{1}{RC_2} \end{bmatrix}$$

$$\begin{bmatrix} \langle v_1\rangle \\ \langle v_2\rangle \end{bmatrix} + \begin{bmatrix} \dfrac{1}{r_sC_1} & \dfrac{2D-1}{2Lf_sC_1}\langle v_2\rangle \\ 0 & \dfrac{-2D+1}{2Lf_sC_2}\langle v_1\rangle \end{bmatrix} \begin{bmatrix} \langle v_s\rangle \\ \langle \Delta d\rangle \end{bmatrix}$$

$$y = \begin{bmatrix} 0 & 1 \end{bmatrix} \begin{bmatrix} v_1 \\ v_2 \end{bmatrix} \quad (8.29)$$

At steady state, the voltage ripple can be expressed as in Equation 8.30. This output voltage ripple is directly affected by the variation of phase-shift duty ratio:

$$\frac{\Delta v_2}{\langle v_2\rangle} = \frac{-2D+1}{4Lf_s^2C_2} \times \frac{\langle v_1\rangle}{\langle v_2\rangle}\Delta d \quad \text{and} \quad \frac{\Delta v_1}{\langle v_1\rangle} = \frac{2D-1}{4Lf_s^2C_1} \times \frac{\langle v_2\rangle}{\langle v_1\rangle}\Delta d \quad (8.30)$$

The transfer function of Equation 8.29 is

$$G_1(s) = \frac{2D(1-D)LfR}{4L^2f^2RC_1C_2r_ss^2 + 4L^2f^2(r_sC_1 + RC_2)s + 4L^2f^2 + D^2Rr_s(D-1)^2}$$

$$G_2(s) = \frac{2(1-2D)V_{in}LfRr_sC_1\left(s + \frac{1}{r_sC_1} - \frac{mD(1-D)}{2LfC_1}\right)}{4L^2f^2RC_1C_2r_ss^2 + 4L^2f^2(r_sC_1 + RC_2)s + 4L^2f^2 + D^2Rr_s(D-1)^2} \quad (8.31)$$

Here

$$v_2(s) = \begin{bmatrix} G_1(s) & G_1(s) \end{bmatrix} \begin{bmatrix} v_s(s) \\ \Delta d(s) \end{bmatrix}$$

The poles of Equation 8.31 are

$$s_{1,2} = \frac{-(r_s C_1 + RC_2) \pm \sqrt{(r_s C_1 + RC_2)^2 - RC_1 C_2 r_s \left(4 + \frac{D^2 Rr_s (D-1)^2}{L^2 f^2}\right)}}{2RC_1 C_2 r_s}$$

(8.32)

If $\Delta d = 0$, then the transient output voltage expression is

$$V_2(s) = G_1(s)v_s(s) = \frac{K}{(s - s_1)(s - s_2)} \frac{v_s}{s}$$

$$= \frac{K_{21}}{(s - s_1)} v_s + \frac{K_{22}}{(s - s_2)} v_s + \frac{K_{23}}{s} v_s \qquad (8.33)$$

After taking the inverse Laplace transformation, the transient output power is

$$P_o = C_2 \frac{dV_2(t)}{dt} V_2(t) + \frac{V_2^2}{R} = (K_{21}e^{s_1 t} + K_{22}e^{s_2 t} + K_{23}v_s)$$

$$\times \left[\left(K_{21}C_2 s_1 + \frac{1}{R} \right) e^{s_1 t} + \left(K_{22}C_2 s_2 + \frac{1}{R} \right) e^{s_2 t} + K_{23}v_s \right] \qquad (8.34)$$

where

$$K = \frac{D(1 - D)}{2LfC_1 C_2 r_s}, \quad K_{21} = \frac{K}{(s_1 - s_2)s_1}, \quad K_{22} = \frac{K}{(s_2 - s_1)s_2}, \quad K_{23} = \frac{K}{s_2 s_1}$$

Similar calculations could be processed if $\Delta d \neq 0$.

Figure 8.20 shows a comparison of simulated results from the above averaged model with results obtained from the model built in MATLAB/Simulink [17]. In Figure 8.20, $\Delta d = 0.002$ is imposed when $0.5\,\text{s} < t < 0.505\,\text{s}$, where $C_1 = 1200\,\mu\text{F}$, $C_2 = 800\,\mu\text{F}$, $nV_1 = V_2 = 600\,\text{V}$, $L = 28\,\mu\text{H}$, $n = 2$, $R = 50\Omega$, $D = 0.01$. It can be seen that the voltage obtained by the averaged model properly tracks the simulated one using MATLAB/Simulink, under steady state as well as transient operation.

The averaged model is a powerful tool to understand the transient processes in power electronic systems. However, the disadvantages are obvious. By averaging the output parameters, the whole system becomes a black box after neglecting the relevant transients with shorter time scales, which are sometimes more important to guarantee safe operation of the system. For example, in the above analysis, if $f_s = 10\,\text{kHz}$, then the timescale of the averaged model is $50\,\mu\text{s}$, half switching period, which defeats our original intention to investigate the microsecond to nanosecond-level processes. Therefore this model is useful to investigate the impact on the output voltage instead of switch current when a variation of duty-ratio occurs.

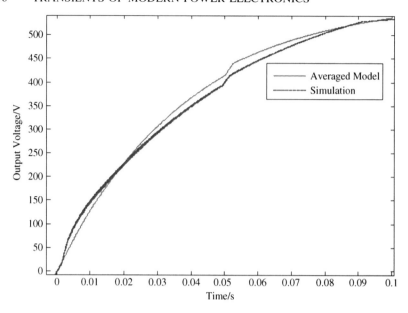

Figure 8.20   Comparison of the calculated output voltage (averaged model vs. simulation using MATLAB/Simulink).

8.5   Summary

Modulated error describes the difference between the intended control performance and the actual one. This error can appear at the microsecond level or minute level. This chapter defines the modulated error from a system perspective. We illustrated three modulated errors in a DC–AC inverter and a DC–DC converter, and demonstrated the methods to analyze, calculate, and mitigate these errors.

In different systems, modulated errors behave differently. A summary of the different system designs shows the necessity of moving the operational condition away from the boundary conditions, such as the regions restrained by the minimum pulse width in the DC–AC inverter and $nV_1 = V_2$ in the DAB DC–DC converter. If the timescale is large enough, the averaged model is a powerful tool for quantitatively evaluating the influence of modulated error.

References

1. Backlund, B. and Carroll, E. (2006) Voltage Ratings of High Power Semiconductors, Product information by ABB Switzerland Ltd, p. 10.
2. Nagel, A., Bernet, S., Bruckner, T. *et al.* (2000) Characterization of IGCTs for series connected operation. Record of IEEE Industry Applications Conference, Vol. 3, pp. 1923–1929.

3. Nagel, A., Bernet, S., Steimer, P.K., and Apeldoorn, O. (2001) A 24 MVA inverter using IGCT series connection for medium voltage applications. Record of the 36th IEEE Industry Application Society Annual Meeting, Vol. 2, pp. 867–870.

4. Kuhn, H. and Schroder, D. (2002) A new validated physically based IGCT model for circuit simulation of snubberless and series operation. *IEEE Transactions on Industrial Applications*, **38** (6), 1606–1612.

5. Ohkami, T., Souda, M., Saito, T. *et al.* (2007) Development of a 40kV series-connected IGBT switch. Proceedings of PCC'07, pp. 1175–1180.

6. Abbate, C., Busatto, G., Fratelli, L. *et al.* (2005) Series connection of high power IGBT modules for traction applications. European Conference on Power Electronics and Applications, pp. 11–14.

7. Wang, H.F., Huang, A.Q., and Wang, F. (2007) Development of a scalable power semiconductor switch (SPSS). *IEEE Transactions on Power Electronics*, **22** (2), 364–373.

8. Bryant, A.T., Wang, Y.L., Finney, S.J. *et al.* (2007) Numerical optimization of an active voltage controller for high-power IGBT converters. *IEEE Transactions on Power Electronics*, **22** (2), 374–383.

9. Consoli, A., Musumeci, S., Oriti, G., and Testa, A. (1995) Active voltage balancement of series IGBTs. Industry Applications Conference, Vol. 3, pp. 2752–2758.

10. Bauer, F., Meysenc, L., and Piazzesi, A. (2005) Suitability and optimization of high-voltage IGBTs for series connection with active voltage clamping. *IEEE Transactions on Power Electronics*, **25** (6), 1244–1253.

11. Bai, H., Zhao, Z., and Liqiang Yuan, E.M. (2007) Optimization design of high-voltage-balancing circuit based on the functional model of IGCT. *IEEE Transactions on Industrial Electronics*, **54** (6), 3012–3021.

12. Liu, W.H., Song, Q., Yu, Q.G., and Chen, Y.H. (2003) Analysis of high di/dt current pulses in three-level NPC inverters using series connected IGCTs and RC snubbers. Industrial Electronics Society Meeting (IECON'03), Vol. 3, pp. 2771–2776.

13. Hua, B. and Zhengming, Z. (2007) Dynamic equivalent circuit design in three-level high voltage inverters based on functional model of IGCT. Applied Power Electronics Conference (IPEC 2007), pp. 1095–1101.

14. Bai, H., Zhao, Z., and Mi, C. (2009) Framework and research methodology of short-timescale pulsed power phenomena in high voltage and high power converters. *IEEE Transactions on Industrial Electronics*, **56** (3), 805–816.

15. Wu, C., Jiang, Q., and Zhang, C. (2005) An optimization method for three-level selective harmonic eliminated pulse width modulation (SHEPWM). International Conference on Electric Machines and Systems, pp. 1346–1350.

16. Bai, H., Mi, C.C., and Gargies, S. (2008) The short-time-scale transient processes in high-voltage and high-power isolated bidirectional DC-DC converters. *IEEE Transactions on Power Electronics*, **23** (6), pp. 2648–2656.

17. Bai, H., Mi, C., Wang, C., and Gargies, S. (2008) The dynamic model and hybrid phase-shift control of a bidirectional dual active bridge DC-DC converter. Industrial Electronics Society Meeting (IECON'08), pp. 2840–2845.

9

Future trends of power electronics

Power electronics is a multidisciplinary subject that involves many research branches including solid state electronics, analogue/digital electronics, microprocessors, digital signal processing, electromagnetics, electric machines, power systems, and so on. Power electronics has been a propeller for many recent industrial paradigm shifts, such as the one currently happening in the automotive industry in moving from conventional gasoline and diesel-based vehicles to electric and plug-in hybrid electric vehicles. In turn, the new industry paradigm creates challenges for power electronics technology which promotes innovation in both.

The future of power electronics will be primarily driven by applications. With more and more power electronics used in power systems, electric and hybrid vehicles, renewable energy systems, and many other industrial, military, and residential applications, innovations in power electronics will focus on boosting efficiency, reducing size, lowering cost, and improving reliability. Such innovations will embrace semiconductor materials, devices, topology, passive components, and packaging technology.

9.1 New materials and devices

Wide-bandgap silicon carbide (SiC) semiconductors have been investigated recently for use in power devices [1–28]. Compared to silicon (Si), SiC has a larger bandgap, a higher electric field breakdown, and a higher thermal conductivity. Thus SiC-based power electronics have the potential for reliable operation at higher junction temperature, higher voltage, higher switching

Transients of Modern Power Electronics, First Edition. Hua Bai and Chris Mi.
© 2011 John Wiley & Sons, Ltd. Published 2011 by John Wiley & Sons, Ltd.

frequency, and thus higher power density than Si devices. These advantages are enabling SiC technology-based power systems to be made smaller, lighter, and more efficient. Higher frequency operation generally can also cut down the size/weight of the system's passive components. High operating temperature allows a larger temperature difference between the heat sink and the cooling media (air or liquid) being used to carry away the heat, which will increase the radiator's effectiveness and decrease its size. Typical Si devices are limited to an operating temperature range up to 125 °C, whereas SiC devices can safely handle temperatures of 200 °C and higher. SiC has the potential for up to a fivefold reduction in converter volume if high-temperature and high-frequency power electronics can be implemented. The reduction in volume will provide increased flexibility in the arrangement of equipment.

Some of the advantages of SiC power devices compared to Si-based power devices can be summarized as follows [1]:

- SiC unipolar devices are thinner and have lower on resistances. At low breakdown voltages ($\sim$50 V), these devices have specific on resistances of 1.12 $\mu\Omega$, approximately 100 times less than the resistances of their Si counterparts. At higher breakdown voltages ($\sim$5000 V), the on resistance of SiC devices goes up to 29.5 mΩ, still much less than the resistance of comparable Si devices. With lower R_{on}, SiC power devices have lower conduction losses; therefore, higher overall converter efficiency is attainable. Greater efficiency will also reduce the need for cooling hardware and lessen the burden of logistics.

- SiC-based power devices have higher breakdown voltages because of their higher electric breakdown fields; for example, Si Schottky diodes typically are commercially available at voltages lower than 300 V, but the first commercial SiC Schottky diodes are already rated at 600 V. Now SiC Schottky diode could reach 1200V@30A.

- SiC has a higher thermal conductivity (4.9 W/cm K for SiC vs. 1.5 W/cm K for Si); therefore, the SiC power device has a lower junction-to-case thermal resistance, R_{th-jc} (0.02 K/W for SiC vs. 0.06 K/W for Si). The device temperature increase is therefore much slower.

- SiC can operate at high temperatures. SiC devices operating at up to 600 °C are mentioned in the literature. Si devices, on the other hand, can operate at a maximum junction temperature of only 150 °C.

- SiC is extremely radiation resistant; that is, radiation does not degrade its electronic properties. Therefore, SiC converters will be useful in aerospace applications because they will reduce the amount of radiation shielding needed, thereby decreasing the weight of the system.

- SiC power devices are more reliable because their forward and reverse recovery characteristics vary only slightly with temperature and time.

- SiC-based bipolar devices have excellent reverse recovery characteristics. With less reverse recovery current, the switching losses and EMI are reduced, and there is less or no need for snubbers. As a result, there may be no need to use soft-switching techniques to reduce the switching losses.

- Because of low switching losses, SiC-based devices can operate at higher frequencies (>200 kHz) that are not possible for Si-based devices at power levels of more than a few tens of kilowatts.

Although there has been significant progress in SiC in the past decade, currently there exist three major technical barriers hampering progress in SiC technologies.

One technical barrier is the very high density of defects in a SiC wafer, which limits the size of the device. SiC devices are currently restricted to small-area devices (less than $10\,\text{mm}^2$). The major defect is the micropipe. These micropipes are tiny holes that are generally clustered. Any device fabricated on an area in which a micropipe breaks the surface will be unlikely to work. In the early 1990s, the density of these micropipes was on the order of 300, but it has been falling since then. It is expected that in the near future SiC wafers could be micropipe-free which makes large-area SiC devices possible.

The second technical barrier to the acceptance of SiC is the lack of suitable device structures that can fit SiC material. Most of the device structures that have been presently tried are copied from Si technologies. Although these structures have been proven in Si, they are not necessarily transferable to SiC. Various problems have been encountered with such structures when they are made from SiC. In the past ten years, unipolar devices have emerged as the structures of choice for low- to medium-voltage (30–3000 V) SiC diodes, and vertical JFETs have emerged as the first commercially feasible SiC switch. Combined with a low-voltage Si MOSFET for ease of driving and to make it a normally off device, the vertical JFET is widely seen as the first SiC voltage-controlled switch to gain acceptance by power electronics designers.

The third technical barrier is the lack of high-temperature, high-power-density packaging techniques for SiC devices. New techniques and technologies must be explored. Currently available packaging techniques are designed for Si devices, which generally have a power density limit of $200\,\text{W/cm}^2$ and a temperature limit of $125\,^\circ\text{C}$, but SiC devices may require a power density of $1000\,\text{W/cm}^2$ and a temperature limit of $300\,^\circ\text{C}$ or more.

Currently there are three major types of SiC wafer commercially available: 4H-SiC, 6H-SiC, and 3C-SiC. Type 6H-SiC is semi-insulating and primarily used for RF devices. Type 4H-SiC is conducting and is primarily used for power device applications.

In the past decade, much effort has been directed toward utilizing proven Si power device structures for SiC technology based on 4H-SiC. Most of the known structures – such as PiN, Schottky, merged PiN–Schottky diodes, bipolar junction transistors, MOSFETs, static induction transistors (SITs), JFETs, thyristors,

GTOs, IGBTs and, more recently, cascode devices – have been designed and implemented in SiC technology. Two types of devices have emerged from the development of these devices:

1. **Unipolar devices:** the functionality of unipolar devices such as Schottky diodes and JFETs has been greatly improved by the use of SiC-based technology. The high electrical breakdown of SiC has enabled these once low-voltage-limited devices to extend their operating range to voltages as high as 3500 V.

2. **Bipolar devices:** bipolar devices have long been the darlings of power electronics engineers because of their low forward voltage drop and ease of control. Devices such as Si IGBTs and IGCTs are widely used in medium- to high-power applications and have gained acceptance through-out the power electronics community. Those devices made of SiC have not been very successful so far because of a number of issues still to be resolved: (i) bipolar degradation or stacking faults; (ii) SiC oxide mobility issues and breakdown; and (iii) lack of a large-area, micropipe-free wafer, and devices.

With bipolar degradation and oxide problems still remaining unresolved, the power device community has turned its attention to unipolar devices and, there-fore, much progress has been made in this area. SiC Schottky diodes rated at up to 3000 V have been demonstrated, and SiC vertical JFETs have been shown to operate reliably at up to 3500 V.

In general, SiC devices are much smaller in physical size compared to Si power devices, resulting in high power densities. A comparison of SiC JFETs and IGBTs is shown in Figure 9.1. The figure indicates that the drift layer thickness (e.g., 20–25 μm) in SiC devices is less than 1/10th the thickness that is required for Si devices (around 300 μm).

The first commercially available SiC devices were Schottky diodes. SiC diodes have been commercialized for a few years. Commercial sources include Cree Inc. (Durham, NC), SiCED (Germany), and SemiSouth (Starkville, MS). As for the design and implementation of SiC switches, only SiC vertical junc-tion field effect transistors (VJFETs) have emerged as commercially feasible. VJFET devices can be obtained from various companies including SiCED and SemiSouth. At the writing of this book, several manufacturers have already made SiC MOSFET and SiC IGBT available as test samples.

Most VJFETS are "normally on" devices, and designers have used them in this mode for power electronics applications. Some other system designers would prefer to use a "normally off" device, and a low-voltage, low-resistance Si MOSFET in a cascode configuration with the SiC VJFET such that a turn-off device can be made. This approach will compromise the benefits of high-temperature operation. Presently, SiC devices have a low current rating, generally less than 15 A. Forming parallel configurations of multiple SiC devices, diodes, and VJFETs is necessary for building SiC power modules. Normally-off SiC

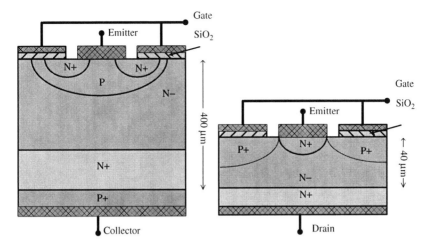

Figure 9.1 A comparison of SiC JFETs and IGBTs. Courtesy Aegis Technology [29].

JFETs are already available from a few manufacturers including SemiSouth and United Silicon Carbide.

Despite the slow progress of development with SiC devices, this technology may ultimately replace Si technology in the future.

Unipolar SiC power devices such as VJFETs and MOSFETs have been shown to reduce losses and consequently improve power conversion efficiency. Normally-off SiC VJFETs avoid the issues associated with the MOSFET SiO_2–SiC interface but must be properly designed to account for the gate channel PN junction. Both normally-on and normally-off VJFETs have been developed by a number of groups. Smaller die-size normally-off VJFETs and 1680 V large-area normally-on devices have been reported. Difficulty in scaling normally-off SiC VJFETs to larger die size has also been reported. SemiSouth has demonstrated the performance of $15 \, mm^2$ 200 V normally-off SiC VJFETs with 114 A saturation current ($950 \, A \, cm^2$) at 25 °C. This represents the highest saturation current reported to date for a normally off SiC VJFET [30].

The SiC VJFET device structure consists of a vertical N channel gated by P^+ regions. The total die area is $15 \, mm^2$ with $11.9 \, mm^2$ active area. Specific on resistance and saturation current density have been normalized to the device active area. Due to the low on resistance of these devices ($28 \, m\Omega$), packaging with minimal parasitic resistance was required.

A photograph of a packaged die is shown in Figure 9.2. DC device characteristics from a single device with nominal threshold voltage ($+1 \, V$) have been measured at case temperatures of 25 and 175 °C. Figure 9.3 shows the output characteristics for $V_{gs} = 2.5$ and 3.0 V. The saturation current decreases from 114 A at 25 °C to 45 A at 175 °C. At 175 °C, the output characteristics at $V_{gs} = 2.5$ and 3.0 V were virtually identical. Figure 9.4 shows the output current

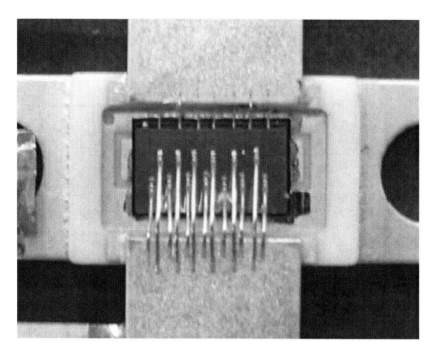

Figure 9.2   Photograph of 15 mm² normally off SiC VJFET. Courtesy SemiSouth Laboratories [30].

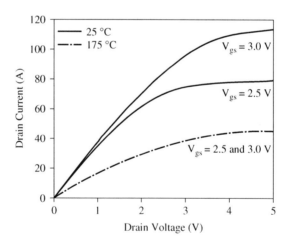

Figure 9.3   Pulsed output characteristics for 1200 V normally off SiC VJFET. Courtesy SemiSouth Laboratories.

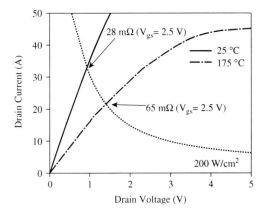

Figure 9.4   Drain current as a function of drain voltage for $V_{gs} = 2.5\,V$. Courtesy SemiSouth Laboratories.

as a function of drain bias for $V_{gs} = 2.5\,\text{V}$. At a power density of $200\,\text{W/cm}^2$, the on resistance increased from $28\,\text{m}\Omega$ at $25\,^\circ\text{C}$ to $65\,\text{m}\Omega$ at $175\,^\circ\text{C}$ (3.3 and $7.8\,\text{m}\Omega\,\text{cm}^2$, respectively).

9.2   Topology, systems, and applications

A power electronic system is expected to realize effective and efficient energy transformation. Due to the limitations of power devices, losses exist as conduction losses, switching losses, and additional losses due to passive components, snubber circuits, and auxiliary circuits. Soft-switching control is utilized to reduce switching loss but is very hard to implement in high-voltage and high-power systems. First, the high-power semiconductor devices have significant variation, which makes soft-switching control (e.g., resonant system) lose the resonant condition. Second, some soft-switching control needs an auxiliary circuit, which needs to be switched with even higher frequency and undergo even higher stress. This may not be feasible or realistic in high-power applications due to highly restricted electrical allowances.

In high-power applications, most of the power electronic converters have an efficiency of 90–98%. In power converters rated with a few hundred kilowatts, even a 2% loss can become difficult for the system to handle the heat generation. Increased heat sink size increases the volume of the whole system. Currently, inverter technology used in hybrid electric vehicles uses $70\,^\circ\text{C}$ coolant that is supplied via a separate cooling loop in the automobile. It is desirable to eliminate the need for an additional cooling loop to reduce cost and complexity in the vehicle. By adopting new materials and devices (SiC, CoolMOS, etc.), creating novel topologies, and adjusting the control algorithm, efficiency of 98% or more can be achieved for many power electronic systems, which is one of the ultimate goals of power electronics.

Take a single-phase power factor correction circuit as an example. In order to reach a high efficiency of 97% or higher, the commonly used methods are: (i) adopt the SiC switches, SiC Schottky diode, or CoolMOS as the switches; (ii) customer design the inductor; and (iii) adjust the switching frequency under different loads, which should take into account the switching loss and iron loss of the transformer, and so on.

Power electronics are very limited in extremely high-voltage, high-power applications. High-voltage DC transmission, for example, 100 kV DC, is one of the future industrial trends. In order to transform the high-voltage direct current to the consumer-end low-voltage alternating current, both a transformer and a DC–AC inverter are essential. However, up to now, most of the power electronic devices cannot meet the demand of high voltage. Even if a 10 kV IGCT is adopted, many modules need to be in series. For a three-level NPC DC–AC inverter, one bridge will need at least 20 IGCTs and 10 diodes. The system becomes bulky and complicated which will decrease the system reliability and increase cost. A fully controllable semiconductor device with higher voltage ratings will be the future direction of power electronics enabling these types of applications.

High switching frequency has been pursued by power electronics engineers. High switching frequency not only represents high operational performance (e.g., low total harmonic distortion (THD)), but also utilizes the down-sized passive components (inductors, transformers, and capacitors). However, high switching frequency of power electronic converters always has limitations. First, it means high switching loss and therefore, we believe, there is a tradeoff between switching frequency and system efficiency. Second, its application directly challenges the present embedded system: 1 MHz switching frequency means 1 μs running time for the core program, which cannot be accomplished by any microcontroller at the present time.

Once the above two conflicts are solved, high-switching-frequency systems will revolutionize power electronic technology. One of the potential applications is wireless power transfer. In 2007 Kurs *et al.* published a paper in *Science* [31] describing the use of four coils in a prototype with power generated by a Colpitts circuit. The source coils and device coils had the same resonant frequency. Using self-resonant coils in a strongly coupled regime, Kurs *et al.* experimentally demonstrated efficient non-radioactive power transfer over distances up to eight times the radius of the coils. In this setup, 60 W with ~40% efficiency over distances of 2 m could be realized.

For the purpose of high-power transfer, convenience of utilization, and environmental protection, coupled magnetic resonance is the primary alternative to realize high-power wireless transfer. The primary concern is how to generate the high-frequency voltage to be applied to the coils because resonance happens at more than 10 MHz as shown in Figure 9.5. A power electronic converter is a promising system to realize effective energy transfer. One possible prototype is shown as Figure 9.6. Power is delivered by the H-bridge. By switching the H-bridge on the primary side, an AC voltage is imposed on the terminals of the

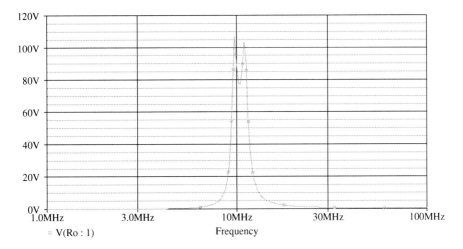

Figure 9.5   Resonant frequencies in wireless power transfer circuits.

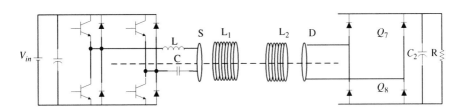

Figure 9.6   The proposed platform for wireless power transfer.

primary coil, S. The secondary coil, D, is connected with a diode-based rectifier, which transforms energy from AC to DC.

The resonant frequency of the coils is beyond 10 MHz. Therefore semiconductors operating in the megahertz range are required. The ceiling switching frequency of commercial MOSFETs is 1 MHz at present. For most Si devices, 1 MHz is in fact far beyond their capability, and whose requirements on the dead band, minimum pulse width, reverse recovery current, and other nonlinear switching processes make 1 MHz switching unrealistic. SiC devices are expected to play a major role in the near future. Even so, generating a megahertz gate-drive signal to drive SiC devices still has a long way to go. Challenges imposed on the microprocessor can also not be neglected.

Lastly, let us take a look at fuel cell vehicles and electric vehicles which are considered as zero-emission vehicles. However, they still depend mainly on the fossil-based sources, for example, carbon and oil to generate electricity and hydrogen. Therefore "zero emissions" will not be 100% true until renewable energy effectively replaces the present fossil-based energy sources.

In this process, power electronic systems will behave as a critical interface between different energy forms as discussed in Chapter 5. Here we look at another example, the high-speed train (HST) system, which has proven to be a safe, comfortable, and efficient method of transportation. Due to its ability to carry a large number of passengers and provide short travel times, the HST has become one of the major tools to alleviate the traffic burden of some of the main traffic corridors in many countries like Japan, France, Germany, Spain, and also recently in South Korea and China. The introduction of the HST is considered to have spatial and socio-economic impacts on regional development. Such improved interregional accessibility leads to a widening of the regional labor market and the establishment of a new corridor economy.

In the development of the HST system, use of solar and wind energy to supply power to the HST can make the system even more environmentally friendly and maximize the economic benefits. The introduction of renewable energy in the HST makes the design of the HST power systems a complicated interaction among power electronic devices, energy integration, process control, and so on. Use of renewable energy in the HST system should combine economy, reliability, high performance, and high comfort simultaneously in order to compete with conventional electric trains. A feasible topology is shown Figure 9.7. Here wind and solar energy stations are distributed along the high-speed railway. A single-phase AC grid supplies power to the HST, collaborating with solar stations, wind stations, batteries, and UCs, which could operate at tens of thousands of volts or even a few hundred thousand volts and thousands of amperes. This will create huge challenges for power electronics.

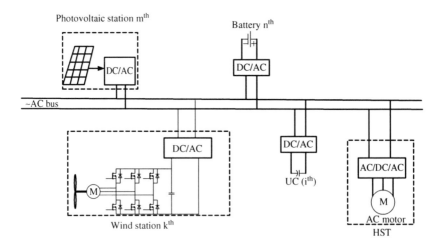

Figure 9.7   Renewable energy-based high-speed rail.

9.3   Passive components

Passive components, including inductors, transformers, and capacitors, play important roles in power electronics-based energy conversion. However, the performance of passive components at present time is inadequate. For example, high-frequency operations of inductors and capacitors often produce harmonics in the megahertz range. This kind of frequency will make a capacitor behave like an inductor and an inductor like a capacitor. The bulky sizes of capacitors and inductors increase the size, weight, and volume of a power converter. The narrow operating temperature range of capacitors limits the applications of power converters in many areas.

Improvements in passive components can be expected to increase the operating temperature range, enhance reliability, reduce size, and lower losses in capacitors, transformers, and inductors:

1. **Thin-film inductor design** [32, 33]: the demand for small and efficient power conversion systems has increased as the market for portable equipment become widespread. The thin-film inductor is one of the promising devices for reducing the size of DC–DC converters. The total thickness of the DC–DC converter IC can be reduced to under a millimeter by using a thin film, which can change magnetic components from a 3D to a 2D structure. This will greatly reduce the dimensions of magnetic components like the transformer and inductor. Traditional inductor design needs windings around the magnetic materials. However, with the development of a soft magnetic material which has low loss, high resistance, high saturate permeability, and high operating frequency, the smaller sized magnetic components can become a reality, as shown in Figure 9.8 [32]. Thin-film inductors will fully excavate the advantages of fast switching devices to minimize the system volume and weight.

2. **Capacitor:** the equivalent series resistance (ESR) and equivalent series inductance (ESL) of the capacitor deteriorate the high-frequency performance of the capacitors. For the snubber capacitor, the ESL will reduce its capability to absorb voltage spikes. For the output capacitor of an EMI filter, the ESL will produce noise when it oscillates at high frequency. Besides the improvements in capacitor technology, commonly used methods are (i) paralleling the capacitors, which increases the volume of the system; and (ii) launching a negative inductance to eliminate the ESL. Yong-sheng *et al.* [34] proposed an ESL eraser, which is laid out on a PCB with two coils, as shown in Figure 9.9. With the aid of the finite element method (FEM), the ESL could be accurately eliminated when $L_M = \text{ESL}$.

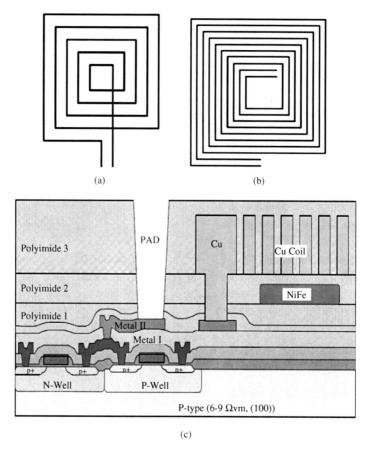

Figure 9.8   Schemes of thin-film windings. (a) thin-film windings and (b) thin-film transformer with coaxial windings. © [2005] IEEE. Reprinted, with permission, from ICEMS 2005. (c) integration of thin-film windings with the power electronics module. Courtesy ETRI.

9.4   Power electronics packaging

The packaging techniques of power electronics involve die attachment, interconnection between die pads and lead frame or substrate, and encapsulation. In the packaging process, some rules need to be followed:

1. Parasitic parameters need to be minimized. This needs to be well addressed in high-switching-frequency applications where safety and efficiency are important.

2. Current capability and insulation are highly demanded simultaneously, which is critical for high-voltage and high-power applications.

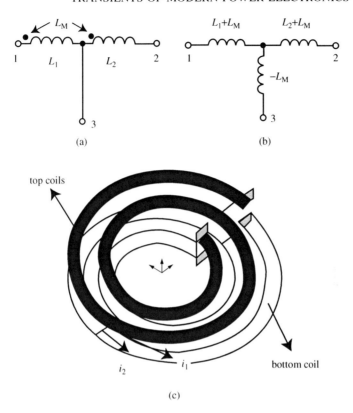

Figure 9.9   Capacitor ESL eraser: (a) scheme of the ESL eraser, (b) equivalent circuit, and (c) real circuit. © [2010] Advanced Technology of Electrical Engineering and Energy.

3. Excellent thermal characteristics are demanded since the integration density is increasing more and more nowadays.

Presently the common packaging technology includes: wire bonding, ribbon bonding, solder joints, conductive epoxy, metal deposition, press pack, sintered silver, and so on. For several decades, the Center of Power Electronic Systems (CPES) in Virginia Tech has represented the innovation frontier of packaging technology for power electronics. CPES milestones in packaging include the metal post interconnect parallel packaging system (MPIPPS) (in 1999), die dimensional ball grid array (DDBGA), flip chip on flex (FCOF) (in 2000), the dimple array, embedded power, pseudo GE-POL, and so on [35]. Figure 9.10 shows the MPIPPS and embedded power module, respectively.

Today more and more packaging tends to further increase the ratio of die area to module area, which significantly reduces the volume of power electronic systems. Another trend in power electronics packaging is to merge the gate-drive

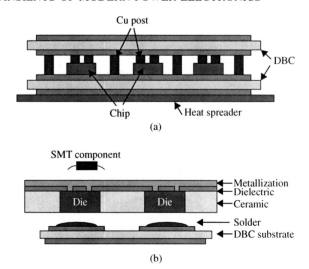

Figure 9.10 Two milestones in packaging technologies by CPES at Virginia Tech: (a) MPIPPS and (b) embedded power module (DBC = Direct Bonded Copper). Courtesy and © Naili Yue.

circuit with the module. IGCT is a typical device which integrates the switch with the gate-drive circuit. For most power electronic modules, semiconductors and gate drives are independent of each other. Even for the IGCT, its gate-drive circuit and GCT are just physically connected. Further effort is expected to embed the gate-drive circuit totally within the module to minimize space and facilitate usage.

Trench gate technology has been employed to reduce the conduction loss of the MOSFET and IGBT, as shown in Figure 9.11 [36]. The thin-wafer punch-through (PT) IBGT has emerged to replace the traditional PT or non-punch-through (NPT) IGBTs with improved performance in terms of conduction loss, turn-off speed, and thermal coefficients as shown in Figure 9.12 [36]. In addition, cooling methods for power electronics system is also evolving. Double-sided cooling and direct cooling of power devices can significantly reduce thermal resistance and increase cooling efficiency. Figure 9.13 shows the trend of estimated chip power loss as a function of chip current density.

9.5 Power line communication

One of the biggest reasons why power electronic converters face obstacles in the power industry is that the grid cannot obtain information on the power electronic system in time. The reliability issues of power electronics systems result in that utility companies can't offer a thorough trust. Power line communication provides a very feasible way to merge power electronics in power systems.

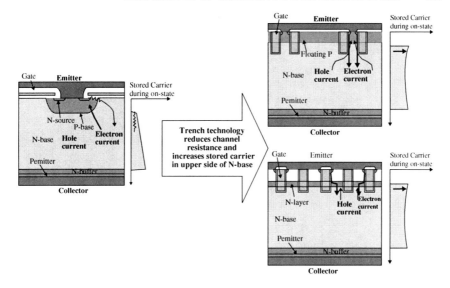

Figure 9.11   Impact of trench gate technology on IGBT conduction loss reduction [36]. © [2007] IEEE. Reprinted, with permission, from the proceedings of the IEEE.

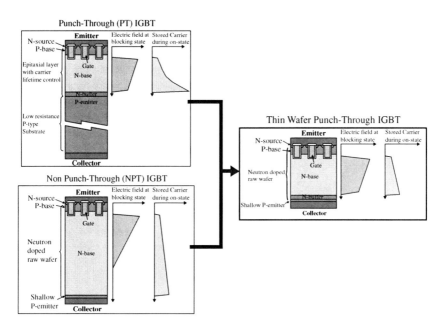

Figure 9.12   Comparison of PT, NPT, and thin-wafer PT IGBTs [36]. © [2007] IEEE. Reprinted, with permission, from the proceedings of the IEEE.

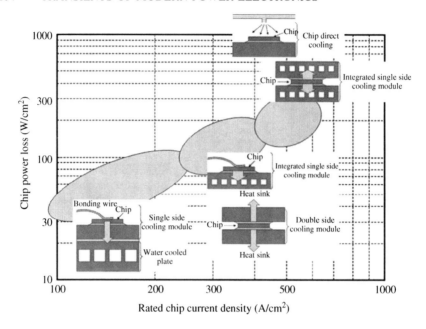

Figure 9.13   Evolution of device packaging technology with regard to cooling efficiency and power density [36]. © [2007] IEEE. Reprinted, with permission, from the proceedings of the IEEE.

For many decades, high-voltage (120 kV and above) transmission lines have been used as the medium for communication between power plants and sub-stations. Also, the power lines are used to communicate between the circuit breakers for the purpose of power system protection. During the last few years this concept has been launched in low-voltage, residential applications where the high-frequency communication signals are coupled with the electrical wiring originally designed for the transmission of electric power, which makes communication between the power electronic system and the power plant fast, reliable, and economic.

Although this seems to be a simple and slick idea, there are many challenges yet to be overcome in order to realize a robust system. When a grid-frequency signal and a high-frequency signal are coupled to the same transmission line, numerous problems might have to be addressed. Separation of the two signals, power signals and communication signals, is one of them and requires good fil-tering. Then, of course, the structure of the transmission line plays a significant role since the natural modes and frequencies of the line are totally dependent on the geometry and materials of the line. The dominant modes and the corre-sponding cutoff frequencies of the line should be carefully determined so that the frequency of the communication signal will be selected appropriately in order to increase the efficiency of the transmission.

The major issues of power line communication will be the selection of the transmission line configuration and the extraction of its parameters using electromagnetic field analysis software. Once the parameters are obtained, a thorough eigenvalue and eigenvector analysis will be necessary to evaluate the dominant modes and the natural frequencies of the transmission line in order to determine the bandwidth of the high-frequency signal to be coupled to the line. The design and implementation of the filters for both the sending end and receiving end of the transmission line is another challenging and important step.

Take a public charging station as an example. When the station begins to charge an electric vehicle, not only will the power be monitored by the power company, but also the information, for example, the potential charging time, vehicle identification, and charging location, will be forwarded to the power company. The advantages are very obvious. First, the power company can coordinate the power flow in time. Second, the communication system is very compact and no extra cables are needed.

9.6   Transients in future power electronics

The transients of power electronics will be more significant when SiC and other new devices replace Si in the future and other new technologies and control algorithms are deployed in the power electronics field. The commutating process will be much shorter, the EMI issue will be more severe, and the stray parameters will play more dominant roles. Therefore, when the devices move to the SiC domain, system reliability will be no less addressed. On the contrary, the transient processes will be more critical.

Besides traditional research on transient behaviors, the transients in future power electronics should also include:

1. **Construction of the system-level SOA:** presently most of the power electronic system design is still based on the component-level SOA from the datasheet. In this case, the control system, sampling system, protection system, and power electronic converters are totally segregated. It is worth to point out that, thermal aspects are a very important issue in power electronics and also very often ignored. Future construction of system-level SOA tends to regard the source, converter, load, and control system in their entirety.

2. **Interaction between the different devices:** when the switches are equipped in the real system, interacting with other components and associated with the control algorithms, they will display variations from the data sheet specifications. Consequently, some unexpected and small-probability pulses will emerge, for example, the sneak pulse in Figure 4.17. When novel semiconductors with faster switching speed come into play, the sneak pulse will appear more frequently, which needs further effort in system design and control.

In summary, power electronics can expect more revolutionary changes in the coming decades. In the future, power electronic devices will increasingly approach ideal switches, but new problems will emerge. How to utilize energy more efficiently and effectively, how to mitigate conflicts due to transient processes, and how to decrease cost and improve reliability, and so on, will still be important topics. A systematic theoretical framework is required, which we believe will be gradually developed with the advancement of power electronic technology.

References

1. Ozpineci, B. (2002) System impact of silicon carbide power electronics on hybrid electric vehicle applications. PhD dissertation. The University of Tennessee.
2. Ozpineci, B., Tolbert, L.M., Islam, S.K., and Hasanuzzaman, M. (2001) Effects of silicon carbide (SiC) power devices on HEV PWM inverter losses. Annual Conference of the Industrial Electronics Society, Vol. 2, pp. 1061–1066.
3. Kelley, R., Mazzola, M.S., and Bondarenko, V. (2006) A scalable SiC device for DC/DC converters in future hybrid electric vehicles. Annual Applied Power Electronics Conference and Exposition, p. 4.
4. Ozpineci, B., Tolbert, L.M., Islam, S.K., and Peng, F.Z. (2002) Testing, characterization, and modeling of SiC diodes for transportation applications. Annual Power Electronics Specialists Conference, pp. 1673–1678.
5. Dreike, P.L., Fleetwood, D.M., King, D.B. *et al.* (1994) An overview of high-temperature electronic device technologies and potential applications. *IEEE Transactions on Components, Packaging, and Manufacturing Technology, Part A,* **17**, 594–609 [See also *IEEE Transactions on Components, Hybrids, and Manufacturing Technology*].
6. Ozpineci, B., Chinthavali, M.S., and Tolbert, L.M. (2005) A 55 kW three-phase automotive traction inverter with SiC Schottky diodes. Vehicle Power and Propulsion Conference, p. 6.
7. Ohashi, H. (2003) Power electronics innovation with next generation advanced power devices. International Telecommunications Energy Conference, pp. 9–13.
8. Zhang, H., Tolbert, L.M., Han, J.H. *et al.* (2010) 18 kW three phase inverter system using hermetically sealed SiC phase-leg power modules. Annual Applied Power Electronics Conference and Exposition, pp. 1108–1112.
9. Richmond, J., Leslie, S., Hull, B. *et al.* (2009) Roadmap for megawatt class power switch modules utilizing large area silicon carbide MOSFETs and JBS diodes. Energy Conversion Congress and Exposition, pp. 106–111.
10. Guy, O.J., Lodzinski, M., Castaing, A. *et al.* Silicon carbide Schottky diodes and MOSFETs: solutions to performance problems. Power Electronics and Motion Control Conference, pp. 2464–2471.
11. Ozpineci, B., Chinthavali, M.S., Tolbert, L.M. *et al.* (2009) A 55-kW three-phase inverter with Si IGBTs and SiC Schottky diodes. *IEEE Transactions on Industry Applications*, **45** (1), 278–285.

12. Chinthavali, M., Tolbert, L.M., Zhang, H. *et al.* (2010) High power SiC modules for HEVs and PHEVs. International Power Electronics Conference, pp. 1842–1848.

13. Pan, S., Mi, C., and Lin, T. (2009) Design and testing of silicon carbide JFETs based inverter. International Power Electronics and Motion Control Conference, pp. 2556–2560.

14. Harada, K., Maki, K., Pounyakhet, S. *et al.* (2010) Switching characteristics of SiC-VJFET and manufacture of inverter. International Power Electronics Conference, pp. 176–179.

15. Burger, B. and Kranzer, D. (2009) Extreme high efficiency PV-power converters. European Conference on Power Electronics and Applications, pp. 1–13.

16. Springmann, D.M., Jahns, T.M., and Lorenz, R.D. (2008) Inverter gate drive and phase leg development for 175 °C operation. Power Electronics Specialists Conference, pp. 2152–2158.

17. Jordan, J., Magraner, J.M., Cases, C. *et al.* Turn on switching losses analysis for Si and SiC diodes in induction heating inverters. European Conference on Power Electronics and Applications, pp. 1–9.

18. Mazzola, M.S. and Kelley, R. (2009) Application of a normally off silicon carbide power JFET in a photovoltaic inverter. Annual Applied Power Electronics Conference and Exposition, pp. 649–652.

19. McBryde, J., Kadavelugu, A., Compton, B. *et al.* (2010) Performance comparison of 1200V silicon and SiC devices for UPS application. IECON 2010 – Annual Conference of IEEE Industrial Electronics Society, pp. 2657–2662.

20. Ritenour, A., Sheridan, D.C., Bondarenko, V., and Casady, J.B. (2010) Saturation current improvement in 1200 V normally-off SiC VJFETs using non-uniform channel doping. International Symposium on Power Semiconductor Devices & ICs (ISPSD 2010), pp. 361–364.

21. Antonopoulos, A., Bangtsson, H., Alakula, M., and Manias, S. (2008) Introducing a silicon carbide inverter for hybrid electric vehicles. Power Electronics Specialists Conference, pp. 1321–1325.

22. Chinthavali, M., Zhang, H, Tolbert, L.M., and Ozpineci, B. (2009) Update on SiC-based inverter technology. Power Electronics Conference, pp. 71–79.

23. Chinthavali, M., Otaduy, P., and Ozpineci, B. (2010) Comparison of Si and SiC inverters for IPM traction drive. Energy Conversion Congress and Exposition, pp. 3360–3365.

24. Araujo, S.V. and Zacharias, P. (2009) Analysis on the potential of silicon carbide MOSFETs and other innovative semiconductor technologies in the photovoltaic branch. European Conference on Power Electronics and Applications, pp. 1–10.

25. Zhang, H. and Tolbert, L.M. (2009) Efficiency of SiC JFET-based inverters. International Conference on Industrial Electronics and Applications, pp. 2056–2059.

26. Sheridan, D.C., Ritenour, A., Kelley, R. *et al.* (2010) Advances in SiC VJFETs for renewable and high-efficiency power electronics applications. International Power Electronics Conference, pp. 3254–3258.

27. Kranzer, D., Wilhelm, C., Reiners, F., and Burger, B. (2009) Application of normally-off SiC-JFETs in photovoltaic inverters. European Conference on Power Electronics and Applications, pp. 1–6.

28. Bergogne, D., Hammoud, A., Tournier, D. *et al.* (2009) Electro-thermal behaviour of a SiC JFET stressed by lightning-induced overvoltages. European Conference on Power Electronics and Applications, pp. 1–8.

29. Aegis Technology, http://www.aegistech.net/.

30. Kelley, R., Stewart, G., Ritenour, A. *et al.* (n.d.) 1700 V Enhancement-mode SiC VJFET for High Voltage Auxiliary Flyback SMPS, SemiSouth Laboratories Inc., http://www.semisouth.com/.

31. Kurs, A., Karalis, A., Moffatt, R. *et al.* (2007) Wireless power transfer via strongly coupled magnetic resonances. *Science*, **317**, 83–86.

32. Ruijun, C., Yugang, Y., and Ying, J (2005) Design of LLC resonant converter with integrated magnetic technology. International Conference on Electrical Machines and Systems, Vol. 2, pp. 1351–1355.

33. Park, I.-Y., Kim, S.G., Koo, J.G. *et al.* (2003) A fully integrated thin-film inductor and its application to a DC-DC converter. *ETRI Journal*, **25** (4), 270–273.

34. Yong-sheng, C.I., Shi-shan, W., Xiao-lin, Z., and Shao-jun, X. (2010) Cancellation of parasitic inductance for filtering capacitor with planar windings. *Advanced Technology of Electrical Engineering and Energy*, **29** (1), 45–58.

35. Yue, N. (2008) Planar packaging and electrical characterization of high temperature SiC power electronic devices. Masters thesis. Virginia Polytechnic and State University.

36. Shen, J.Z. and Omura, I. (2007) Power semiconductor devices for hybrid, electric, and fuel cell vehicles. *Proceedings of the IEEE*, **95** (4), 778–789.

Index

Transients of Modern Power Electronics, First Edition. Hua Bai and Chris Mi.
© 2011 John Wiley & Sons, Ltd. Published 2011 by John Wiley & Sons, Ltd.

CPSIA information can be obtained at www.ICGtesting.com
Printed in the USA
BVOW040637120911

270233BV00004B/8/P

9 780470 686645